Lecture Notes in Computer Science   12709

More information about this subseries at http://www.springer.com/series/7407

Alexander Raschke · Dominique Méry (Eds.)

Rigorous State-Based Methods

8th International Conference, ABZ 2021
Ulm, Germany, June 9–11, 2021
Proceedings

Editors
Alexander Raschke ⓘ
Ulm University
Ulm, Germany

Dominique Méry ⓘ
University of Lorraine
Vandœuvre-lès-Nancy, France

ISSN 0302-9743   ISSN 1611-3349   (electronic)
Lecture Notes in Computer Science
ISBN 978-3-030-77542-1   ISBN 978-3-030-77543-8   (eBook)
https://doi.org/10.1007/978-3-030-77543-8

LNCS Sublibrary: SL1 – Theoretical Computer Science and General Issues

This Springer imprint is published by the registered company Springer Nature Switzerland AG
The registered company address is: Gewerbestrasse 11, 6330 Cham, Switzerland

Preface

The International Conference on Rigorous State-Based Methods (ABZ 2021) is an international forum for the cross-fertilization of related state-based and machine-based formal methods, mainly Abstract State Machines (ASM), Alloy, B, TLA+, VDM, and Z. Rigorous state-based methods share a common conceptual foundation and are widely used in both academia and industry for the design and analysis of hardware and software systems.

The name ABZ was invented at the first conference held in London in 2008, where the ASM, B, and Z conference series merged into a single event. The second conference, ABZ 2010, was held in Orford, Canada, where the Alloy community joined the event; ABZ 2012 was held in Pisa, Italy, which saw the inclusion of the VDM community in the conference series (but not in the title); and ABZ 2014 was held in Toulouse, France, which brought the inclusion of the TLA+ community. Lastly, ABZ 2016 was held in Linz, Austria, and ABZ 2018 in Southampton, UK. In 2018 the Steering Committee decided to retain the (well-known) acronym ABZ and add the subtitle "International Conference on Rigorous State-Based Methods" to make more explicit the intention to include all state-based formal methods.

In 2020, the ABZ conference should have been held in Ulm, Germany. Unfortunately, the conference had to be cancelled at short notice due to the worldwide rampant COVID-19 virus and was postponed to this year, with the hope of welcoming all participants personally in Ulm. Unfortunately this hope did not come true, so we had to organize the conference as a virtual event after all. Since the proceedings were ready before the cancellation of the 2020 conference, we decided to publish them immediately. We also launched a new call for papers for ABZ 2021 so that researchers had the opportunity to publish new results in a timely manner.

Because the ABZ conference is normally hosted every two years we had not defined a new case study, and with the restrictions of the pandemic and the associated increased efforts, for example for teaching, being felt in the research community, significantly fewer papers were submitted to ABZ 2021. There were 15 papers submitted from authors in 8 countries spread over Europe, Asia, and America. Fortunately, the submitted papers were of very high quality, so that the four reviews per paper were consistently positive and only one paper had to be rejected. A total of 6 full research papers and 8 short research papers were accepted. All accepted papers cover broad research areas on theoretical, systems, or practical aspects of state-based methods.

The doctoral symposium, which was organized for the first time in 2020, also took place in 2021. Three PhD students submitted a four-page abstract describing their research projects and received constructive feedback from the senior researchers of the ABZ community. Each of the submitted abstracts was also evaluated by a separate Program Committee.

The conference was due to be held during June 9–11, 2021, in Ulm, Germany, but unfortunately the successes of the COVID-19 vaccination program will probably not be seen for several months, so the conference was held virtually. In addition to the new submissions, the authors of ABZ 2020 were also invited to present their papers, which fortunately many took advantage of and thus a comprehensive program could be put together.

Unfortunately, due to consequences of the COVID-19 crisis in the personal environment, one of the keynote speakers understandably had to cancel at short notice. However, we were all the more pleased to listen to the keynotes of Ana Cavalcanti, University of York, UK, on "RoStar technology — a roboticist's toolbox for combined proof and sound simulation" and Gilles Dowek, Inria/ENS Paris-Saclay, France, on "Sharing proofs across logics and systems: a boost for formal methods?"

The EasyChair conference management system was set up for ABZ 2021, supporting submissions and the review process.

We would like to thank all the authors who submitted their work to ABZ 2021. We are grateful to the Program Committee members for their high-quality reviews and discussions. Finally, we wish to thank the Organizing Committee members for their continuous support.

We hope the vaccination program will also reach poor countries as fast as possible such that the COVID-19 crisis will decrease within the next two years and we can meet together in person at ABZ 2023 in France.

For readers of these proceedings, we hope these papers are interesting and they inspire ideas for future research that can be presented at the next ABZ!

April 2021   Alexander Raschke
                                                   Dominique Méry

Organization

General Chairs

Alexander Raschke   Ulm University, Germany
Dominique Méry   Université de Lorraine, LORIA, France

Program Committee

Yamine Ait Ameur   IRIT/INPT-ENSEEIHT, France
Paolo Arcaini   National Institute of Informatics, Japan
Richard Banach   University of Manchester, UK
Egon Boerger   Università di Pisa, Italy
Eerke Boiten   De Montfort University, UK
Michael Butler   University of Southampton, UK
Andrew Butterfield   Trinity College Dublin, Ireland
David Deharbe   ClearSy System Engineering, France
Juergen Dingel   Queen's University, Canada
Flavio Ferrarotti   Software Competence Centre Hagenberg, Austria
Mamoun Filali-Amine   IRIT, France
Marc Frappier   Université de Sherbrooke, Canada
Angelo Gargantini   University of Bergamo, Italy
Vincenzo Gervasi   University of Pisa, Italy
Gudmund Grov   Norwegian Defence Research Establishment (FFI), Norway
Stefan Hallerstede   Aarhus University, Denmark
Klaus Havelund   Jet Propulsion Laboratory, USA
Ian J. Hayes   The University of Queensland, Australia
Thai Son Hoang   University of Southampton, UK
Frank Houdek   Daimler AG, Germany
Alexei Iliasov   Newcastle University, UK
Felix Kossak   Software Competence Center Hagenberg, Austria
Regine Laleau   Paris-Est Créteil University, France
Thierry Lecomte   ClearSy, France
Michael Leuschel   University of Düsseldorf, Germany
Alexei Lisitsa   University of Liverpool, UK
Atif Mashkoor   Johannes Kepler University, Austria
Jackson Mayo   Sandia National Laboratories, USA
Stephan Merz   Inria Nancy, France
Stefan Mitsch   Carnegie Mellon University, USA
Rosemary Monahan   Maynooth University, Ireland
Mohamed Mosbah   University of Bordeaux, France
Cesar Munoz   NASA, USA

Contents

Regular Research Articles

Unbounded Barrier-Synchronized Concurrent ASMs for Effective
MapReduce Processing on Streams 3
  Zilinghan Li, Shilan He, Yiqing Du, Senén González,
  and Klaus-Dieter Schewe

Towards ASM-Based Automated Formal Verification
of Security Protocols 17
  Chiara Braghin, Mario Lilli, and Elvinia Riccobene

Verifying System-Level Security of a Smart Ballot Box 34
  Dana Dghaym, Thai Son Hoang, Michael Butler, Runshan Hu,
  Leonardo Aniello, and Vladimiro Sassone

Proving the Safety of a Sliding Window Protocol with Event-B 50
  Sophie Coudert

Event-B Formalization of Event-B Contexts 66
  Jean-Paul Bodeveix and Mamoun Filali

Validation of Formal Models by Timed Probabilistic Simulation 81
  Fabian Vu, Michael Leuschel, and Atif Mashkoor

Short Articles

Sterling: A Web-Based Visualizer for Relational Modeling Languages 99
  Tristan Dyer and John Baugh

Extending ASMETA with Time Features 105
  Andrea Bombarda, Silvia Bonfanti, Angelo Gargantini,
  and Elvinia Riccobene

About the Concolic Execution and Symbolic ASM Function Promotion
in CASM 112
  Philipp Paulweber, Jakob Moosbrugger, and Uwe Zdun

Towards Refinement of Unbounded Parallelism in ASMs Using
Concurrency and Reflection 118
  Fengqing Jiang, Neng Xiong, Xinyu Lian, Senén González,
  and Klaus-Dieter Schewe

The CamilleX Framework for the Rodin Platform   124
   Thai Son Hoang, Colin Snook, Dana Dghaym, Asieh Salehi Fathabadi,
   and Michael Butler

Extensible Record Structures in Event-B   130
   Asieh Salehi Fathabadi, Colin Snook, Thai Son Hoang, Dana Dghaym,
   and Michael Butler

Formalizing and Analyzing System Requirements of Automatic Train
Operation over ETCS Using Event-B   137
   Robert Eschbach

Automatic Transformation of SysML Model to Event-B Model for Railway
CCS Application.   143
   Shubhangi Salunkhe, Randolf Berglehner, and Abdul Rasheeq

Short Articles of the PhD-Symposium (Work in Progress)

Formal Meta Engineering Event-B: Extension and Reasoning
The *EB4EB* Framework.   153
   Peter Riviere

A Modeling and Verification Framework for Security Protocols   158
   Mario Lilli

Formalizing the Institution for Event-B in the Coq Proof Assistant   162
   Conor Reynolds

Author Index   167

Regular Research Articles

Unbounded Barrier-Synchronized Concurrent ASMs for Effective MapReduce Processing on Streams

Zilinghan Li[1], Shilan He[1], Yiqing Du[1], Senén González[2],
and Klaus-Dieter Schewe[1]([✉])

[1] UIUC Institute, Zhejiang University, Haining, China
{zilinghan.18,shilan.18,yiqing.18,kd.schewe}@intl.zju.edu.cn
[2] P&T Connected, Linz, Austria

Abstract. MapReduce supports the processing of large data sets in parallel. It has been shown that MapReduce is an example for the use of the *bulk synchronous parallel* (BSP) bridging model, a model for parallel computation on a fixed set of processors comprising alternating computation and communication phases. In this article we extend the normal execution of MapReduce from processing large finite data sets to processing stream queries with input data stream assumed to continue indefinitely. We classify stream queries into three classes, *memoryless*, *semi-memoryless* and *memorable*, and provide the model for each class using MapReduce based on BSP. In addition, as some stream queries require large amounts of computing sources, the BSP computation model is extended to a model with unbounded many agents, but preserving the barrier synchronization. A behavioral theory is developed for this model extending the behavioral theory of the BSP model. This comprises an axiomatization, the definition of Infinite-Agent BSP abstract state machines (Inf-Ag-BSP-ASM) and the proof that such ASMs capture the unbounded synchronized computations. Finally, we show how MapReduce processing can be further improved on grounds of the unbounded extension.

Keywords: MapReduce · Stream query · Infinite-agent BSP model · Behavioral theory · Concurrent ASM

1 Introduction

MapReduce provides a programming model for processing large data sets in an asynchronous and distributed way [6]. A MapReduce computation comprises a *map* phase processing input data in an asynchronous and parallel way to obtain intermediate key-value pairs, a *shuffle* phase redistributing the data and a *reduce* phase aggregating intermediate key-value pairs to yield the final results.

The *bulk synchronous parallel* (BSP) bridging model [16] is a widely-used model for parallel computations by a fixed number of agents. A BSP computation consists of a sequence of *supersteps*, and each *superstep* is composed of a

© Springer Nature Switzerland AG 2021
A. Raschke and D. Méry (Eds.): ABZ 2021, LNCS 12709, pp. 3–16, 2021.
https://doi.org/10.1007/978-3-030-77543-8_1

computation phase and a communication phase. In a computation phase each agent works independently without any form of interaction until it completes the local computation. When all agents have completed their local computations, they continue with a communication phase to exchange data. With all agents having completed their communication, they return to a new computation phase and thus begin a new *superstep*. The BSP model has many useful applications as demonstrated among others in [5,10,13,14].

As explained in [9] BSP computations are specific concurrent algorithms that are characterised by a fixed number of agents, the alternation of computation and communication phases and the barrier synchronization. Consequently, they are captured by specific concurrent abstract state machines (concurrent ASMs), the BSP-ASMs, as shown by the behavioral theory of BSP computations.

There exists a strong connection between MapReduce and BSP model: for the *map* phase of MapReduce each agent executes the user-defined map function without interaction, which indicates that the *map* phase corresponds to a computation phase in the BSP model. The *shuffle* phase corresponds to a communication phase in the BSP model, in which agents interact with each other by exchanging data. For the *reduce* phase agents execute the user-defined reduce function again without interaction, so it also corresponds to a computation phase. In addition, as MapReduce algorithms proceed, some of the agents may be executing a map function, while others are executing a reduce function in the same computation phase. The strong connection indicates that MapReduce can be captured by the BSP model, which was already sketched in [9]. More examples on how MapReduce is realized based on grounds of the BSP model can be found in [14].

A general assumption underlying MapReduce is that while the input data sets can be large, they are still bounded. In this article we investigate an extension based on the BSP model enabling the handling of stream queries, where the input data stream is assumed to continue indefinitely. Then we also extend the BSP bridging model following the behavioral theory to further support MapReduce query processing.

We first classify the stream queries according to their concatenation property and divide them into three different classes: *memoryless*, *semi-memoryless* and *memorable*. In a nutshell, a memoryless stream process can forget the input data once they have been processed, a semi-memoryless stream process has to keep some aggregation of the data for the processing of following data, and a memorable stream process requires all processed data to be kept. Then we provide the models for these different classes of stream queries using MapReduce based on the BSP model.

As the data stream is assumed to continue indefinitely, also the agents executing MapReduce are dedicated to processing the stream query indefinitely. However, we may want to allow some of the agents to *leave* the computation and other new agents to *join* the computation. In particular, some stream queries require a large amount of computation time and a large memory space for data storage, especially for some *semi-memoryless* and *memorable* queries, so it will be advantageous not to have any limit on the number of agents. Theoretically, we may permit an infinite number of agents, but following the theory of concurrent algorithms [3] only finitely many of them can be active in each *superstep*. Nonetheless, both the *join/leave* behavior and an unbounded number of agents are beyond the expressiveness of general BSP model, where the number of agents is fixed and bounded.

To extend BSP computation and allow such behaviors, we develop an extended behavioral theory for Infinite-Agent BSP (Inf-Ag-BSP) computations, which permits changing numbers of agents in all supersteps, but preserves the barrier synchronization. The notion of *behavioral theory* was coined to characterise the method used in the sequential ASM thesis characterising sequential algorithms in a language-independent way [11]. This work therefore became the first behavioral theory that was investigated. The same method was later used for recursive algorithms [4], synchronous parallel algorithms [1,2,7,8,15], and asynchronous concurrent algorithms [3]. As BSP computations are special concurrent algorithms, a specialised behavioral theory was also developed for the BSP model [9]. In our extended behavioral theory we will exploit that concurrent ASMs can deal with an infinite number of agents [3].

2   BSP-ASMs for Stream Queries

In our extension of MapReduce to handle stream queries, we restrict ourselves to *abstract computable* stream queries following the definitions and notations in [12]. A stream query $Q : Stream \to Stream$ is *abstract computable* if there exists a function $K : finStream \to finStream$ such that the result of Q can be obtained by concatenating the results of K applied to larger and larger prefixes of the input, i.e.

$$Q : \mathbf{s} \mapsto \bigodot_{k=0}^{size(\mathbf{s})} K(\mathbf{s}^{\leq k}), \tag{1}$$

where we use $\odot$ to denote concatenation[1]. Here K is called the *kernel* of Q. In the following K is always used to denote the *kernel* of the corresponding stream query.

According to Eq. 1, evaluating stream query Q on stream $\mathbf{s}$ requires evaluating K on the whole stream. However, when handling stream queries via MapReduce, streams are split into several pieces that are evaluated independently. Then the individual results are aggregated to yield the final result. In the following stream queries are classified into three classes based on their concatenation property. We show how each class can be realised via MapReduce based on the BSP model.

---

[1] The concatenation ($\odot$) used here is not same as the common concatenation denoted by $\sum$. It works more like aggregation and its real functionality varies among different scenarios, but we still use the term concatenation to be consistent with [12].

2.1   Memoryless Stream Queries

In a *memoryless* stream query the result of new coming input stream is independent of the previous input streams, and the output of a larger stream can be obtained by direct concatenation of the outputs of its smaller substreams. This gives rise to the following definition.

Definition 1. A stream query $\mathcal{Q}$ is said to be *memoryless* if $\mathcal{Q}$ can defined as follows:

$$\mathcal{Q} : \mathbf{s} \mapsto \bigodot_{k=0}^{size(\mathbf{s})} K(\mathbf{s}^{\leq k}) = \bigodot_{k=0}^{size(\mathbf{s})} K(s_k) \qquad (2)$$

where s_k is the kth element of stream $\mathbf{s}$.

Example 1. Consider the query $\mathcal{Q}_1$ which returns an output stream containing all numbers greater than a threshold value t_x. When $\mathbf{s}$ is divided into several substreams, then we can filter out the values greater than t_x in each substream and concatenate them together to yield the final outputs. Since the output can be obtained by direct concatenation of the outputs of smaller substreams, $\mathcal{Q}_1$ is *memoryless*.

MapReduce Model. For every MapReduce cycle i, we extract a substream $\mathbf{s}^{(i)}$ from the input stream and execute the *map* function on this substream to evaluate $K(\mathbf{s}^{(i)})$; the result is concatenated to the previous result in the *reduce* phase in the following way:

$$\mathcal{Q}(\mathbf{s}) = \bigodot_{k=1}^{i} K(\mathbf{s}^{(k)}) \text{ with } \mathbf{s} = \sum_{k=1}^{i} \mathbf{s}^{(k)} \qquad (3)$$

The following ASM 1 gives an ASM rule for agent j handling a *memoryless* stream query. The variable $task_j$ is the current phase of agent j which can be either map or reduce. The rule $bsp\_sync()$ is used for synchronization which indicates the end of the current *superstep* for an agent. The agent will not continue with the next *superstep* until all agents complete their current *superstep*. The rule $bsp\_send()$ can send the local data to other agents and the sent data is only available in the next *superstep*, i.e. after $bsp\_sync()$. The rule $bsp\_get()$ fetches the data sent by other agents in the previous *superstep*.

ASM 1 (MapReduce for *Memoryless* Stream Query based on BSP).

    IF $task_j =$ "map"
    THEN $map\_out_j := Map\_Function(\mathbf{s}_j)$
               $bsp\_send(map\_out_j)$
               $bsp\_sync()$
    IF $task_j =$ "reduce"
    THEN $map\_out_j := bsp\_get()$
               $reduce\_out_j := concat(reduce\_out_j, map\_out_j)$
               $bsp\_sync()$

2.2 Semi-memoryless Stream Queries

For a *semi-memoryless* stream query it is not sufficient to yield the output of a larger stream by simply concatenating the outputs of smaller substreams s_i. Some further information of these smaller substreams, denoted as $\mathbf{I}(s_i)$, is also required to compute the output. This leads to the following definition.

Definition 2. A stream query $\mathcal{Q}$ is said to be *semi-memoryless* if $\mathcal{Q}$ can be defined as follows:

$$\mathcal{Q} : s \rightarrow \mathbf{F}_{agg}(K(s_1), K(s_2), \mathbf{I}(s_1), \mathbf{I}(s_2)) \text{ with } s = s_1 + s_2, \tag{4}$$

where K is the kernel of $\mathcal{Q}$, $\mathbf{I}(s_i)$ is a set of informative variables of substreams s_i, and $\mathbf{F}_{agg}$ is the function that aggregates the outputs of smaller substreams and yields the output of the larger stream.

Example 2. Consider the query $\mathcal{Q}_2$ which returns the average value of the numbers arrived so far. When evaluating the average of stream s, where s is divided into two substreams s_1 and s_2, first execute function K which calculates the average value of s_1 and s_2 respectively. In this example, the lengths of two substreams are also needed to calculate the total average, namely, $\mathbf{I}(s_i) = \{len(s_i)\}$. Therefore, $\mathcal{Q}_2$ is *semi-memoryless* and can be evaluated as follows:

$$\begin{aligned}
\mathcal{Q}(s) &= \mathbf{F}_{agg}(K(s_1), K(s_2), \mathbf{I}(s_1), \mathbf{I}(s_2)) \\
&= \mathbf{F}_{agg}(avg(s_1), avg(s_2), \{len(s_1)\}, \{len(s_2)\}) \\
&= avg(s_1) \cdot \frac{len(s_1)}{len(s_1) + len(s_2)} + avg(s_2) \cdot \frac{len(s_2)}{len(s_1) + len(s_2)}
\end{aligned} \tag{5}$$

MapReduce Model. For every MapReduce cycle i we extract a substream $s^{(i)}$ from the input stream and execute *map* function on this substream to evaluate both $K(s^{(i)})$ and the informative set $\mathbf{I}(s^{(i)})$, then aggregate with previous results in the *reduce* phase in the following way:

$$\mathcal{Q}(s) = \mathbf{F}_{agg}(K(\sum_{k=1}^{i-1} s^{(k)}), K(s^{(i)}), \mathbf{I}(\sum_{k=1}^{i-1} s^{(k)}), \mathbf{I}(s^{(i)})) \text{ with } s = \sum_{k=1}^{i} s^{(k)} \tag{6}$$

The following ASM 2 gives an ASM rule for agent j handling a *semi-memoryless* stream query. The term *tag* is used to denote the informative set $\mathbf{I}$. The function *bsp_send_tag()* sends the *tag* to other agents and the function *bsp_get_tag()* fetches the *tags* sent by others. The variable *local_tag$_j$* is the previously stored *tags* of agent j (i.e. $\mathbf{I}(\sum_{k=1}^{i-1} s^{(k)})$) in Eq. 6). The function *Agg* plays the role of $\mathbf{F}_{agg}$ which aggregates the previous and current tags and map-results.

ASM 2 (MapReduce for *Semi-memoryless* Stream Query based on BSP).

> **IF** $task_j$ = "map"
> **THEN** $map\_out_j := Map\_Function(\mathbf{s}_j)$
> $tag\_out_j := Tag\_Fuction(\mathbf{s}_j)$
> $bsp\_send(map\_out_j)$
> $bsp\_send\_tag(tag\_out_j)$
> $bsp\_sync()$
> **IF** $task_j$ = "reduce"
> **THEN** $map\_out_j := bsp\_get()$
> $tag\_out_j := bsp\_get\_tag()$
> $reduce\_out_j := Agg(reduce\_out_j, map\_out_j, local\_tag_j, tag\_out_j)$
> $bsp\_sync()$

2.3   Memorable Stream Queries

The following definition of a *memorable* stream query indicates that a large set of information, or even the whole input stream **s** is required to obtain the output of the stream query.

Definition 3. A stream query $\mathcal{Q}$ is said to be *memorable* if $\mathcal{Q}$ is semi-memoryless as defined by Eq. 4 and the cardinality of $\mathbf{I}(\mathbf{s}_1)$ and $\mathbf{I}(\mathbf{s}_2)$ is in the Θ-Class[2] of the cardinality of $\mathbf{s}_1$ and $\mathbf{s}_2$, respectively, i.e.

$$|\mathbf{I}(\mathbf{s}_1)| \in \Theta(|\mathbf{s}_1|) \text{ and } |\mathbf{I}(\mathbf{s}_2)| \in \Theta(|\mathbf{s}_2|) \tag{7}$$

or equivalently,

$$|\mathbf{I}(\mathbf{s}_1) + \mathbf{I}(\mathbf{s}_2)| \in \Theta(|\mathbf{s}|) \tag{8}$$

Example 3. Consider the query $\mathcal{Q}_3$ which returns the median value of the numbers arrived so far. Given any stream, every number can be the candidate of the median, which means the whole stream **s** is required to determine the median. When evaluating the median of **s**, where **s** is divided into two substreams $\mathbf{s}_1$ and $\mathbf{s}_2$, we have $\mathbf{I}(\mathbf{s}_1) = \mathbf{s}_1$ and $\mathbf{I}(\mathbf{s}_2) = \mathbf{s}_2$. Therefore, $\mathcal{Q}_3$ is *memorable*.

MapReduce Model. According to Definition 3 a *memorable* stream query is a special case of a *semi-memoryless* stream query. Hence, the MapReduce model of *memorable* query is almost the same as the one given by Algorithm 2. One of the most important features of *memorable* query is that the cardinality of the informative set is unbounded when the input stream continues indefinitely, and infinite storage space and computation time is required in the aggregation step.

Though in real-life practice no stream will be infinite and a stream query for an almost-infinitely long input stream is rarely of interest, the *memorable*

---

[2] Θ-Class is the intersection of O-Class and Ω-Class which provides an asymptotically tight bound for functions.

stream query raises the problem how to handle stream queries that require a large amount of computations and a large storage space. A feasible solution is to have a large amount of available agents[3], which allows agents to *join*, e.g. when there are many parallel computation threads, or *leave*, e.g. when the computation is almost sequential. This gives sufficient computing sources and storage space to guarantee the stream query evaluation to succeed. However, neither an unbounded number of agents nor a dynamic growth and shrinking of a set of agents is foreseen in the BSP model. It is therefore required to extend the BSP model to Infinite-Agent BSP model as we will do in the following section.

3 An Extended Unbounded BSP Model

We now extend the BSP model such that an unbounded number of agents is permitted, though only finitely many of them can be simultaneously active. In addition, the extended BSP model should allow agents to *join* or *leave* the BSP computation. The extended BSP model will be called Infinite-Agent BSP (Inf-Ag-BSP) computation model. In the following subsections a behavioral theory of Inf-Ag-BSP computation will be developed, which extends the behavioral theory of the general BSP computation.

In general, a behavioral theory consists of an axiomatization of a class of algorithms, a definition of an abstract machine model and a proof that the abstract machine model captures the class of algorithms stipulated by the axiomatization. The axiomatization is essentially a language-independent definition of the algorithm class, and it is given by a set of postulates. Then the characterization theorem contains two parts: (1) instances of the abstract machine model satisfiy the postulates in the axiomatization; (2) every algorithm defined by the postulates can by step-by-step simulated by an abstract machine, i.e. an instance of the abstract machine model.

3.1 Axiomatization

There are two essential differences between the general BSP computation model and the extended Inf-Ag-BSP computation model: (1) The total number of available agents can be countably infinite; (2) The agents can *join* or *leave* the set of active agents in the BSP computation. Therefore, the axiomatization of Inf-Ag-BSP can be derived from the axiomatization of the general BSP computation model developed in [9] with those two differences taken into consideration.

For the first difference, we just slightly modify the number of available agents from finite to countably infinite. This extension is feasible, as general BSP computations are just restricted concurrent algorithms, and as such they can have an infinite number of agents [3]. As for the second difference we introduce a local variable $join_i$ and a joining set $\mathcal{J}_n$. The variable $join_i$ is a function symbol of

---

[3] In theory, we can assume that the number is countably infinite, provided we restrict the model such that only finitely many of them will be simultaneously active.

arity 0 in the local signature of each algorithm. When the algorithm is going to *join* the set of active agents, the variable will be set to **true**; when the algorithm decides to *leave* the BSP, it will be reset to **false**. The joining set $\mathcal{J}_n$ contains the identifiers i of the active agents which have joined in state S_n. In addition, requiring $\mathcal{J}_n$ to be a finite set ensures that only finitely many agents join at the same time.

However, there remains another issue concerning the *join/leave* behavior, which concerns when an agent is allowed to *join* or *leave* the BSP computation. As for *join* we require any new agent to *join* only in the communication phase because it needs to receive data to manipulate in the next computation phase. As for *leave* we require that an agent can *leave* only after communication, i.e. in the computation phase, because an agent may send some computed data to other agents in the communication phase. All above considerations motivate the axiomatization of Inf-Ag-BSP in the following Definition 4, which originates from that of the general BSP model with several necessary restrictions and modifications.

Definition 4. An *Infinite-Agent BSP (Inf-Ag-BSP) algorithm* is a concurrent algorithm $\mathcal{B} = \{(a_i, \mathcal{A}_i) \mid i \in \mathbb{N}\}$ that satisfies the Inf-Ag-BSP signature restriction and Inf-Ag-BSP communication-and-joining postulate defined in the following two definitions.

Definition 5 (Inf-Ag-BSP signature restriction). A concurrent algorithm $\mathcal{B} = \{(a_i, \mathcal{A}_i) | i \in \mathbb{N}\}$ *satisfies the Inf-Ag-BSP signature restriction* iff the signature of $\mathcal{A}_0$ is $\Sigma_0 = \{bar_i, join_i \mid i \in \mathbb{N}\backslash\{0\}\} \cup \{barrier\}$, and the signature of any other $\mathcal{A}_i$ $(i \in \mathbb{N}\backslash\{0\})$ is $\Sigma_i = \Sigma_{i,loc} \cup \{c_{i,j}, c_{j,i} \mid j \in \mathbb{N}\backslash\{0\}, j \neq i\} \cup \{barrier\}$ such that the following conditions are satisfied:

(1) The subsignatures $\Sigma_{i,loc}$ are pairwise disjoint, and each of them contains the corresponding function symbols bar_i and $join_i$ of arity 0;
(2) For $\mathcal{A}_i$, all locations $(c_{i,j}, \boldsymbol{v})$ are write-only;
(3) For $\mathcal{A}_i$, all locations $(c_{j,i}, \boldsymbol{v})$ are read-only;
(4) The function symbol $barrier$ has arity 0 and is monitored for all agents a_i $(i \neq 0)$;
(5) For $\mathcal{A}_0$, it monitors bar_i and $join_i$.

Definition 6 (Inf-Ag-BSP communication-and-joining postulate[4]). For a concurrent algorithm $\mathcal{B} = \{(a_i, \mathcal{A}_i) \mid i \in \mathbb{N}\}$ with *concurrent run* $S_0, S_1, \cdots$, update sets $\Delta_n = \bigcup_{a \in Ag_n} \Delta_a(res(S_{a(n)}, \Sigma_a))$ for $S_{n+1} = S_n + \Delta_n$, and a joining set $\mathcal{J}_n = \{i \mid \mathrm{val}_{S_{a_i(n)}}(join_i) = \mathbf{true}\}$, $\mathcal{B}$ *satisfies the Inf-Ag-BSP communication-and-joining postulate* iff the following conditions are satisfied:

(1) The cardinality of joining set $\mathcal{J}_n$ is finite for all n;
(2) If $i \notin \mathcal{J}_n$, then either $a_i \notin Ag_n$ or $(join_i, \mathbf{true}) \in \Delta_n$;

---

[4] The definitions of the notations used here can be found in [9, Def. 2.2].

(3) If there exists an update $(l, v) \in \Delta_n$ with a location of form $l = (c_{i,j}, v)$, then all updates in Δ_n(except $(bar_x,$ **false**) and $\{(bar_y, \textbf{true}), (join_y, \textbf{true})\}$) have this form and $val_{S_{a(n)}}(barrier) = \textbf{true}$ holds for all $a \in Ag_n$;

(4) If there exists no update $(l, v) \in \Delta_n$ with a location of the form $l = (c_{i,j}, v)$, then $val_{S_{a(n)}}(barrier) = \textbf{false}$ holds for all $a \in Ag_n$;

(5) If non-trivial update $(join_i, \textbf{true}) \in \Delta_n$, then $(bar_i, \textbf{true}) \in \Delta_n$;

(6) If non-trivial update $(join_i, \textbf{false}) \in \Delta_n$, then $val_{S_{a_i(n)}}(bar_i) = \textbf{false}$;

(7) Whenever $(bar_i \wedge \neg barrier) \vee (\neg bar_i \wedge barrier)$ holds for any agent $i \in \mathcal{J}_n$ in state S_n, then $\Delta_{a_i(n)}(res(S_n, \Sigma_i)) = \emptyset$;

(8) The location $barrier$ is lazy, namely, it only changes value when all bar_i for $i \in \mathcal{J}_n$ have changed their values from **true** to **false** (or **false** to **true**).

The postulates in Definition 6 is used to describe the barrier-synchronization and $join/leave$ behaviors of Inf-Ag-BSP algorithm. The first condition in the postulate requires that only finitely many agents can be active at the same time. Actually, the finiteness of $\mathcal{J}_n$ follows from the finiteness of the sets Ag_n, so condition (1) can be derived from the others. The second condition states that agents a_i with $val_{S_{a_i(n)}}(join_i) = \textbf{false}$ cannot contribute to an update in state S_n except by an attempt to $join$ the BSP computation in next state S_{n+1}. The third and forth separate update sets into only two cases: one is the update sets corresponding to the communication phase and another is the update sets corresponding to the computation phase. The fifth and sixth condition specify when the value of $join_i$ can be changed, i.e. when can algorithms $join$ or $leave$ the BSP algorithm: an algorithm must $join$ and first enter the communication phase and must $leave$ in the computation phase[5]. The seventh condition states that all algorithms involved in the BSP algorithm (i.e. $val_{S_{a_i(n)}}(join_i) = \textbf{true}$) have to wait for $barrier$ to become **true** (or **false**, respectively) after setting bar_i to **true** (or **false** respectively).

Analogously to the theory of BSP algorithms we can show concurrent runs of Infinite-Agent BSP algorithms are indeed organised in supersteps, each comprising a computation phase, in which the individual algorithms $\mathcal{A}_i$ operate only on their local locations, followed by a communication phase, in which the individual algorithms $\mathcal{A}_i$ operate only on their channel locations. The proof is analogous to the corresponding proof in [9, Thm. 2.1].

Theorem 1 (Superstep Separation Theorem). *Let $S_0, S_1, \ldots$ be a concurrent run of an Infinite-Agent BSP algorithm $\mathcal{B} = \{(a_i, \mathcal{A}_i) \mid 0 \leq i \leq k\}$. Then there exists a sequence $0 = i_0 < i_1 < \ldots$ with the following properties:*

1. *The value of barrier in state S_{i_j} is false for even j and true for odd j.*
2. *For all x with $i_j \leq x < i_{j+1}$ the update set Δ_x (with $S_x + \Delta_x = S_{x+1}$) satisfies:*

---

[5] Note that it still has to be ensured that an agent leaving the computation does this after completing its step. This, however, has to be ensured by the specification of the programs of the agents.

(a) If j is even, then all updates in Δ_x affect locations of the form $(f, \boldsymbol{v})$ with $f \in \bigcup_{i=1}^{k} \Sigma_{i,loc}$.

(b) If j is odd, then none of the updates in Δ_x affects a location of the form $(f, \boldsymbol{v})$ with $f \in \bigcup_{i=1}^{k} \Sigma_{i,loc}$.

3. For all j, all x with $i_{j-1} \le x < i_j$ and all agents $a \in Ag_x$ we have $i_{j-1} \le a(x)$.

3.2   Infinite-Agent BSP Abstract State Machine

As Inf-Ag-BSP algorithms define a restricted class of concurrent algorithms, an abstract machine model for Inf-Ag-BSP algorithms can also be defined as restricted concurrent ASM, which we will call Infinite-Agent BSP abstract state machine (Inf-Ag-BSP-ASM). Definition 7 below defines the rules for such ASM.

Definition 7 (Rules). For $i \in \mathbb{N}\backslash\{0\}$ let $\Sigma_i = \Sigma_{i,loc} \cup \{c_{i,j}, c_{j,i} \mid j \in \mathbb{N}\backslash\{0\}, j \ne i\} \cup \{barrier\}$ be signatures such that the subsignatures $\Sigma_{i,loc}$ are pairwise disjoint and contain 0-ary function symbols bar_i and $join_i$. Then the *process rules*, *barrier rules* and *join rules* over Σ_i are defined as follows:

(1) Each assignment rule $f(t_1, \ldots, t_n) := t_0$ with $f \in \Sigma_{i,loc}\backslash\{bar_i, join_i\}$ and terms $t_1, \ldots, t_n$, where n is the arity of f, is a process rule. Assignment rules $bar_i := \textbf{true}$ and $join_i := \textbf{false}$ are also process rules.

(2) Each assignment rule $c_{i,j}(t_1, \ldots, t_n) := t_0$ with $j \in \mathbb{N}\backslash\{0\}, j \ne i$ and terms $t_1, \ldots, t_n$, where n is the arity of $c_{i,j}$, is a barrier rule. Assignment rule $bar_i := \textbf{false}$ is also a barrier rule.

(3) The only join rule is the parallel composition of $join_i := \textbf{true}$ and $bar_i := \textbf{true}$.

(4) If all $r_1, r_2, \cdots, r_m$ are process rules (or barrier rules, respectively), then the parallel composition $(r_1 \mid r_2 \mid \cdots \mid r_m)$ is also a process rule (or barrier rule, respectively).

(5) If r is a process rule (or barrier rule, respectively) and φ is a Boolean term, then (IF φ THEN r) is also a process rule (or barrier rule, respectively).

An *Inf-Ag-BSP rule* over Σ_i ($i \in \mathbb{N}\backslash\{0\}$) is a rule of the form:

$$(\text{IF } join_i \wedge \neg bar_i \wedge \neg barrier \text{ THEN } r_{i,proc}) \mid$$
$$(\text{IF } join_i \wedge bar_i \wedge barrier \text{ THEN } r_{i,comm}) \mid$$
$$(\text{IF } \neg join_i \text{ THENCHOOSE } r_{i,join} \text{ OR } skip)$$

with a process rule $r_{i,proc}$, a barrier rule $r_{i,comm}$ and the join rule $r_{i,join}$.
Let $\Sigma_0 = \{bar_i, join_i \mid i \in \mathbb{N}\backslash\{0\}\} \cup \{barrier\}$ and $\mathcal{J} = \{i \mid \text{val}(join_i) = \textbf{true}\}$ be the joining set, a *switch rule* is a rule over signature Σ_0 of the form:

$$\text{SWITCH} \equiv$$
$$\left(\text{IF } \neg barrier \wedge \bigwedge_{i \in \mathcal{J}} bar_i \text{ THEN } barrier := \textbf{true}\right) \mid$$
$$\left(\text{IF } barrier \wedge \bigwedge_{i \in \mathcal{J}} \neg bar_i \text{ THEN } barrier := \textbf{false}\right)$$

For machines executing an Inf-Ag-BSP algorithm each active machine, i.e. val($join_i$) = **true**, has to execute its local process rule $r_{i,proc}$ until the local variable bar_i is set to **true**. Once the "synchronisor" machine $\mathcal{M}_0$ updates $barrier$ to **true** according to the value of bar_i of all involved machines, $\mathcal{M}_i$ has to execute its barrier rule $r_{i,comm}$, until the local variable bar_i is reset to **false**. Active machines can only *leave* during a computation phase, so $join_i :=$ **false** is a process rule. An inactive machine can either do nothing or *join* and then enter the communication phase. These considerations lead to the following definition.

Definition 8. An *Infinite-Agent BSP abstract state machine* (Inf-Ag-BSP-ASM) is a concurrent ASM $\{(a_i, \mathcal{M}_i) | i \in \mathbb{N}\}$, where the signatures Σ_i of machines $\mathcal{M}_i$ are defined as in Definition 7, the rule of $\mathcal{M}_0$ is SWITCH over Σ_0, the rule of other $\mathcal{M}_i$ ($i \neq 0$) is an InfAg-BSP rule over Σ_i, and the joining set $\mathcal{J} = \{i \mid \text{val}(join_i) = \textbf{true}\}$ is always finite.

3.3 Characterization Theorem

The characterization theorem shows that Inf-Ag-BSP-ASM captures the Inf-Ag-BSP algorithms. As mentioned before the proof consists of two parts: (1) Inf-Ag-BSP-ASM satisfies the postulates for Inf-Ag-BSP algorithms in the axiomatization; (2) every Inf-Ag-BSP algorithm defined by the postulates can be step-by-step simulated by an Inf-Ag-BSP-ASM.

Lemma 1. *Inf-Ag-BSP-ASM satisfies the postulates for Inf-Ag-BSP algorithms in the axiomatization.*

Proof. Let $\mathcal{M} = \{(a_i, \mathcal{M}_i) | i \in \mathbb{N}\}$ be an Inf-Ag-BSP-ASM. According to Definition 8, the signature of $\mathcal{M}$ satisfies the Inf-Ag-BSP signature restriction. It remains to show that $\mathcal{M}$ also satisfies the Inf-Ag-BSP communication-and-joining postulate.

Condition (1) is trivially satisfied because $\mathcal{J}$ is required to be always finite in the definition of $\mathcal{M}$. Condition (2) is a consequence of the Inf-Ag-BSP rule. When val($join_i$) = **false**, the first two if-conditions of the Inf-Ag-BSP rule cannot be satisfied and the third if-condition must be satisfied. Then the machine either does nothing or sets both $join_i$ and bar_i to **true**, which matches what Condition (2) requires. For Conditions (3) and (4) we can directly exploit the proof in [9].

For Condition (5) the update ($join_i$, **true**) is non-trivial, i.e. the initial value of $join_i$ is **false**. Therefore, it corresponds to the join rule $r_{i,join}$. According to the definition of join rules we must have also $bar_i :=$ **true** $\in r_{i,join}$, which implies Condition (5). For Condition (6) we exploit that the update ($join_i$, **false**) must come from a process rule. This can only be executed, if val(bar_i) = **false** holds, so Condition (6) is satisfied.

For Condition (7), $(bar_i \wedge \neg barrier) \vee (\neg bar_i \wedge barrier)$ for $i \in \mathcal{J}_n$ is equivalent to $(join_i \wedge bar_i \wedge \neg barrier) \vee (join_i \wedge \neg bar_i \wedge barrier)$. In this case none of the three if-conditions of Inf-Ag-BSP rule are satisfied, so the ASM does nothing and the update set is $\emptyset$. Condition (8) is again a direct and trivial consequence of the definition of the rule SWITCH.

Lemma 2. *Every Inf-Ag-BSP algorithm defined by the postulates can be step-by-step simulated by an Inf-Ag-BSP-ASM.*

Proof. Assume that $\mathcal{B} = \{(a_i, \mathcal{A}_i) | i \in \mathbb{N}\}$ is an Inf-Ag-BSP algorithm. Given that Inf-Ag-BSP algorithms define a restricted class of concurrent algorithms, we let our abstract machine be a cASM $\mathcal{M} = \{(a_i, \mathcal{A}_i) | i \in \mathbb{N}\}$ with the same signature and the same concurrent runs as $\mathcal{B}$, which is the result of the behavioral theory of concurrent algorithm in [3].

The machine $\mathcal{M}_0$ of agent 0 executes a SWITCH rule, which is used for controlling the separation of the computation phase and communication phase of BSP algorithm. The rules for other machines $\mathcal{M}_i$ have the form

$$(\text{IF } \varphi_1 \text{ THEN } r_{i,1}) \mid \cdots \mid (\text{IF } \varphi_m \text{ THEN } r_{i,m})$$

We can exploit the same argument as in [9] to show that the rules r_i can be written in the form of Inf-Ag-BSP rule when replacing r_i by the following rule:

$$(\text{IF } join_i \wedge \neg barrier \wedge \neg bar_i \text{ THEN } r_{i,proc}) \mid$$
$$(\text{IF } join_i \wedge barrier \wedge bar_i \text{ THEN } r_{i,comm}) \mid$$
$$(\text{IF } \neg join_i \text{ THENCHOOSE } r_{i,join} \text{ OR } skip)$$

with rules $r_{i,proc}, r_{i,comm}, r_{i,join}$ defined as follows:

$$r_{i,proc} = (\text{IF } \varphi_1 \text{ THEN } r_{i,1}) \mid \cdots \mid$$
$$(\text{IF } \varphi_{m'} \text{ THEN } r_{i,m'})$$
$$r_{i,comm} = (\text{IF } \varphi_{m'+1} \text{ THEN } r_{i,m'+1}) \mid \cdots \mid$$
$$(\text{IF } \varphi_{m''} \text{ THEN } r_{i,m''})$$
$$r_{i,join} = (\text{IF } \varphi_{m''+1} \text{ THEN } r_{i,m''+1}) \mid \cdots \mid$$
$$(\text{IF } \varphi_m \text{ THEN } r_{i,m})$$

which completes the proof that $\mathcal{B}$ can be simulated by an Inf-Ag-BSP-ASM.

Combining Lemmata 1 and 2 gives the following claimed result.

Theorem 2 (Characterization theorem). *Infinite-Agent BSP abstract state machine captures Infinite-Agent BSP algorithms.*

4   Processing of Stream Queries with MapReduce Using Inf-Ag-BSP ASMs

After extending BSP model to the Inf-Ag-BSP model let us revisit our MapReduce models for *memoryless* and *semi-memoryless* stream queries in the ASMs 1 and 2. Based on the Inf-Ag-BSP model we refine the previous ASMs, which leads to Algorithms 3 and 4. The main difference is that this time we check whether an agent has *joined* the family of active agents of the BSP computation or an agent has *left* before each iteration. The rule *Join_Update()* updates the value of $join_i$ at the begining of the computation phase of each superstep. The refined ASMs allow agents to *join* or to *leave*.

ASM 3 (MapReduce for *Memoryless* Stream Query based on Inf-Ag-BSP)

IF $join_j = $ **true**
THEN $join_j := Join\_Update()$
              // Update the value of $join_i$ in current superstep
         IF $join_j = $ **true**
         IF $task_j = $ "map"
         THEN $map\_out_j := Map\_Function(s_j)$
                    $bsp\_send(map\_out_j)$
                    $bsp\_sync()$
         IF $task_j = $ "reduce"
                    $map\_out_j := bsp\_get()$
                    $reduce\_out_j := concat(reduce\_out_j, map\_out_j)$
                    $bsp\_sync()$

ASM 4 (MapReduce for *Semi-memoryless* Stream Query based on Inf-Ag-BSP)

IF $join_j = $ **true**
THEN $join_j := Join\_Update()$
              // Update the value of $join_i$ in current superstep
         IF $join_j = $ **true**
         IF $task_j = $ "map"
                    $map\_out_j := Map\_Function(s_j)$
                    $tag\_out_j := Tag\_Fuction(s_j)$
                    $bsp\_send(map\_out_j)$
                    $bsp\_send\_tag(tag\_out_j)$
                    $bsp\_sync()$
         IF $task_j = $ "reduce"
                    $map\_out_j := bsp\_get()$
                    $tag\_out_j := bsp\_get\_tag()$
                    $reduce\_out_j := Aggre(reduce\_out_j, map\_out_j,$
                                               $local\_tag_j, tag\_out_j)$
                    $bsp\_sync()$

5   Conclusion

In this paper, we extended MapReduce to handle different classes of stream queries based on the BSP model. We distinguished memoryless, semi-memoryless and memorable stream queries. The increasing need for memory and the increasing number of parallel *map/reduce* tasks in memorable stream queries motivated the extension of the general BSP model to Inf-Ag-BSP model. For this we developed a behavioral theory for the Inf-Ag-BSP model based on the theory of the general BSP model, and we showed how the extended model is used to refine the processing of stream queries. Furthermore, the Inf-Ag-BSP model is not bound to MapReduce, so it can also be employed to handle other parallel computation problems, especially those requiring a large amount of computation work.

References

1. Blass, A., Gurevich, Y.: Abstract state machines capture parallel algorithms. ACM Trans. Comput. Logic **4**(4), 578–651 (2003)
2. Blass, A., Gurevich, Y.: Abstract state machines capture parallel algorithms: correction and extension. ACM Trans. Comp. Logic **9**(3), 1–32 (2008)
3. Börger, E., Schewe, K.-D.: Concurrent abstract state machines. Acta Inf. **53**(5), 469–492 (2015). https://doi.org/10.1007/s00236-015-0249-7
4. Börger, E., Schewe, K.D.: A behavioural theory of recursive algorithms. Fundam. Inf. **177**(1), 1–37 (2020)
5. Costa, V.G., Marín, M.: A parallel search engine with BSP. In: Third Latin American Web Congress (LA-Web 2005), pp. 259–268. IEEE Computer Society (2005). https://doi.org/10.1109/LAWEB.2005.7
6. Dean, J., Ghemawat, S.: MapReduce: simplified data processing on large clusters. In: Proceedings of the 6th Conference on Symposium on Opearting Systems Design & Implementation, OSDI 2004, vol. 6, p. 10. USENIX Association (2004). http://dl.acm.org/citation.cfm?id=1251254.1251264
7. Dershowitz, N., Falkovich-Derzhavetz, E.: On the parallel computation thesis. Logic J. IGPL **24**(3), 346–374 (2016). https://doi.org/10.1093/jigpal/jzw008
8. Ferrarotti, F., Schewe, K.D., Tec, L., Wang, Q.: A new thesis concerning synchronised parallel computing - simplified parallel ASM thesis. Theor. Comp. Sci. **649**, 25–53 (2016). https://doi.org/10.1016/j.tcs.2016.08.013
9. Ferrarotti, F., González, S., Schewe, K.D.: BSP abstract state machines capture bulk synchronous parallel computations. Sci. Comput. Program. **184**, 102319 (2019). https://doi.org/10.1016/j.scico.2019.102319
10. Gava, F., Pommereau, F., Guedj, M.: A BSP algorithm for on-the-fly checking CTL* formulas on security protocols. J. Supercomput. **69**(2), 629–672 (2014). https://doi.org/10.1007/s11227-014-1099-8
11. Gurevich, Y.: Sequential abstract-state machines capture sequential algorithms. ACM Trans. Comp. Logic **1**(1), 77–111 (2000). https://doi.org/10.1145/343369.343384
12. Gurevich, Y., Leinders, D., Van den Bussche, J.: A theory of stream queries. In: Arenas, M., Schwartzbach, M.I. (eds.) DBPL 2007. LNCS, vol. 4797, pp. 153–168. Springer, Heidelberg (2007). https://doi.org/10.1007/978-3-540-75987-4_11
13. Inda, M.A., Bisseling, R.H.: A simple and efficient parallel FFT algorithm using the BSP model. Parallel Comput. **27**(14), 1847–1878 (2001)
14. Pace, M.F.: BSP vs. MapReduce. In: Ali, H.H., et al. (eds.) Proceedings of the International Conference on Computational Science (ICCS 2012). Procedia Computer Science, vol. 9, pp. 246–255. Elsevier (2012)
15. Schewe, K.-D., Wang, Q.: A simplified parallel ASM thesis. In: Derrick, J., et al. (eds.) ABZ 2012. LNCS, vol. 7316, pp. 341–344. Springer, Heidelberg (2012). https://doi.org/10.1007/978-3-642-30885-7_27
16. Valiant, L.G.: A bridging model for parallel computation. Commun. ACM **33**(8), 103–111 (1990). https://doi.org/10.1145/79173.79181

Towards ASM-Based Automated Formal Verification of Security Protocols

Chiara Braghin⑩, Mario Lilli⑩, and Elvinia Riccobene⁽✉⁾⑩

Computer Science Department, Università degli Studi di Milano,
via Celoria 18, Milan, Italy
{chiara.braghin,mario.lilli,elvinia.riccobene}@unimi.it

Abstract. In the security protocols domain, formal verification is more and more highly demanded to guarantee security assurance: humans increasingly depend on the use of connected devices in their daily life, so they must be protected against possible threats and accidents. However, formal verification, and in general the use of formal methods, is slowed by myths and misconceptions, mainly due to their mathematical base, which discourages many designers or engineers from their adoption.

In this paper, we pose the basis for the long-term development of an ASM-based user-friendly framework for the formal verification of security protocols. We introduce a mathematical-based set of templates to formalise common patterns in security protocols and a set of security properties. These templates facilitate the protocol formal verification by providing built-in functions and domains, as well as transition rules and property schema, to be customised according to the specific protocol to be verified. The effectiveness of this approach is shown by means of their application to a number of well-known cryptographic security protocols.

Keywords: Cryptographic protocols · Security assurance · Abstract state machines

1 Introduction

With the rise of Internet and other open networks, the world became more and more connected, thus technology and data have taken more significant roles in our daily lives. A report conducted in 2017 by Domo [20] estimated that by 2020 (without taking in consideration the COVID-19 outbreak that intensified the process) 1.7MB of data would be created every second for every person on earth. This massive amount of data, often sensitive, is shared among users by means of communication protocols. A large number of security protocols have been developed and deployed in order to provide security guarantees (such as authentication of actors, or secrecy of some pieces of information) through the

---

The work was partially supported by the SEED Project SENTINEL.

A. Raschke and D. Méry (Eds.): ABZ 2021, LNCS 12709, pp. 17–33, 2021.
https://doi.org/10.1007/978-3-030-77543-8_2

application of cryptographic primitives. However, the design of security protocols, despite their apparent simplicity, is particularly error-prone. Many communication protocols do not use up-to-date security features or implement them wrongly, and it is difficult to detect these small vulnerabilities without a formal analysis. Moreover, security errors cannot be detected by functional software testing because they appear only in the presence of a malicious adversary. For this reason, many published protocols have been found flawed many years after they have been implemented and used in real applications. Thus, formal verification of security protocols has become a key issue. It is also important to enforce the security during the design phase obtaining the so called *security by design*. Indeed, flaws found after the design phase are costly to be patched, and the cost increases exponentially during the next development phase [23].

Research in the field of protocol verification has been very active in the last thirty years. It reached a fairly mature state, however most of the tools are not widely adopted by industry as one would expect. Unfortunately, the adoption of formal methods is slowed by myths and misconceptions, mainly due to their mathematical base, which discourages many designers or engineers [18]. The common problems of these tools are: a difficult modelling language, making the writing of the model error-prone as well; a verification process that might require user interaction and knowledge of the tool's internal; the verification results difficult to interpret and to bind to the original protocol. For this reason, many of the protocols verified so far have been a case study for a specific tool and each time a new protocol has to be verified, the modelling of the protocol has to start from scratch, although often the protocols share the same structure.

In this paper, we want to tackle the problem from a little bit different perspective: we want to reduce the gap between software engineers or designers' background and formal methods notation. To this aim, as a first step, we introduce a mathematical-based set of templates to formalise common patterns in security protocols and a set of security properties. These templates facilitate the protocol formal verification by providing built-in functions and domains, as well as transition rules and property schema, to be customised according to the specific protocol to be verified. Specifically, we propose a library of functions and domains that can be used to develop Abstract State Machine (ASM) [14] models of security protocols. ASM specifications are instances of a model template that reflects the common structure usually shared by the security protocols. We also provide a set of schema for temporal logic formulas that specify common security goals of cryptographic protocols. If instantiated according to the specific protocol information, resulting properties can be automatically verified by the AsmetaSMV model checker [2]. The modelling formalism that we have defined, although not yet near to natural language, has the advantage of being understood as "pseudocode over abstract data", so it is easy to read and use. Our approach has been applied to well-known cryptographic security protocols and to the scenario of IoT protocols [25].

The results presented here are the first step toward a long term research project. In the future, we want to hide the mathematical formalism behind a high customisable user interface. More precisely, we want to formalise a notation that links univocally a formalised characteristic of a secure communication protocol with a graphical element. Final users can easily drag and drop the components that they need to build their protocol model. Moreover, we want to make available in our framework the automatic implementation of security protocols by exploiting techniques that map ASM models to code (e.g., to C++ [12] or Java [3]).

The rest of the paper is organised as follows. In Sect. 2 we briefly recall some basic definitions on security protocols and on the ASM formal method; we also present two protocols used in the paper as running case studies. Section 3 introduces the library of domains and functions to model the various elements of a security cryptographic protocol and to model the concept of the attacker; we also show the application of the library primitives to model the two running examples; finally, we present formula schemas for security protocol verification and we instantiate them for the two protocol examples. Section 4 relates our work with existing approaches, and Sect. 5 concludes the paper.

2 Background

2.1 Security Protocols

Although there is a wide range of protocols, differing by the number of principals (or actors), the number of messages, the goals of the protocol (that may often be expressed with a list of desired security properties), they all share a common structure. Indeed, a communication protocol consists of a sequence of messages between two or more principals. Each message may be written in the form:

$$\text{M1. } A \rightarrow B : message\_payload$$

which specifies:

- The *principals* (or *actors*) exchanging messages (in general, symbols A and B represent arbitrary principals, S a server). In particular, the direction of the arrow specifies the sender and the receiver of the message.
- The order in which messages are sent, and their specific *payload*. In particular, M1 is a label identifying the message, whereas *message_payload* specifies the actual content of the message.

In secure protocols, payloads can be partially or totally ciphered, either by symmetric-key encryption (in this case, K_{AB} is often used to specify a key shared between actors A and B, used both to encrypt and decrypt), or by asymmetric-key encryption (here, K_B and K_B^{-1} are used to specify a public and private key of B, respectively). Message payloads can contain other information, such as nonces (N), timestamps (T), etc.

The security goals are often defined with respect to CIA (Confidentiality, Integrity, Authentication) triad. The most common are *confidentiality* or *integrity* of message payloads, or *entity authentication* (i.e., the process by which one entity is assured of the identity of a second entity that is participating to the same session of a protocol, thus, they share the same values of the protocol parameters, such as session keys, nonces, etc.).

Consider for example two classic protocols (that are still in use in a revised version) that will be used throughout the paper as running examples to introduce also the ASM notation. The Needham-Schroeder public-key protocol (NSPK, forshort) has been introduced in 1978 for mutual authentication (here, we omit

Needham-Schroeder public-key protocol	SSL protocol
M1. $A \rightarrow B : \{A, N_A\}_{K_B}$	M1. $A \rightarrow B : \{K_{AB}\}_{K_B}$
M2. $B \rightarrow A : \{N_A, N_B\}_{K_A}$	M2. $B \rightarrow A : \{N_B\}_{K_{AB}}$
M3. $A \rightarrow B : \{N_B\}_{K_B}$	M3. $A \rightarrow B : \{C_A, \{N_B\}_{K_A^{-1}}\}_{K_{AB}}$

the exchanges with the certification authority to get the public keys). It consists of three messages: in the first message, principal A sends to B a message containing her identity, A, and a nonce, N_A, to avoid replay attacks (i.e., reuse of old messages, often called a *challenge* message), that only B can decrypt with his private key. B's answer (message M2) is ciphered with A's public key and contains nonce N_A to authenticate B (he is the only one able to decrypt message M1 and obtain N_A in clear), and a nonce N_B to authenticate A with B. Since message M2 is encrypted with A's public key, she is the only one who can decrypt it, thus if B receives message M3 containing nonce N_B encrypted with his public key, A is authenticated, too.

The SSL protocol was introduced by Netscape to exchange a session key and mutual authentication. In message M1, client A sends a session key K_{AB} to server B. In the second message, B produces a challenge N_B, which A signs and returns to B along with a certificate C_A, in case B does not have her public key. Also in this case, A is authenticated when B receives the signed nonce.

In general, a secure protocol is used to communicate along an *insecure* network in an *untrusted* environment, such as Internet. That is, the communication channel is shared among possibly untrusted principals. This means that any arbitrary adversary E may try to subvert a protocol run. It is common to distinguish between *passive* and *active* attackers, which differ in the capabilities they have. The passive attacker is able to eavesdrop the channel, whereas an active attacker controls the network and can delete, inject, modify and intercept any message and is only limited by the constraints of the cryptographic methods used (i.e., the classical symbolic Dolev-Yao model [19] assuming perfect cryptography). In this fully untrusted scenario, it is also wrong to assume that an intended principal of the protocol always behaves honestly.

Both the above protocols are vulnerable to this attack. In NSPK, if A starts a session of the protocol with a dishonest principal E, E may start a new session of the protocol with B to whom (s)he forwards all the messages (s)he receives from A, where (s)he just changes the key used to encrypt. As a consequence, B

will think to run a protocol session with A and not with E (the solution is to add the identity of the intended receiver in message M2, i.e., $\{N_A, N_B, B\}_{K_A}$).

In the SSL protocol, B can be a dishonest principal opening a session with another principal C, making C to believe (s)he is talking to A (i.e., B impersonates A with C by showing the nonce signed by A in message M3). The attack may be avoided by modifying message M3 in $\{C_A, \{A, B, K_{AB}, N_B\}_{K_A^{-1}}\}_{K_{AB}}$.

2.2 Abstract State Machines in a Nutshell

ASMs [13,14] is the formalism that we use for modeling security protocols. They are a state-based formal method, which extends Finite State Machines (FSMs) by replacing unstructured control *states* with algebraic structures (i.e., domains of objects with functions defined on them), and performing state transitions by firing *transition rules*. At each computation step, all transition rules are executed in parallel by leading to simultaneous (consistent) updates of a number of locations – i.e., memory units defined as pairs (*function-name, list-of-parameter-values*)–, and therefore changing functions interpretation from one state to the next one. Location *updates* are given as assignments of the form $loc := v$, where loc is a location and v its new value. Among other *rule constructors*, those used for our purposes are constructors for: guarded updates (`if-then`, `switch-case`), parallel updates (`par`), sequential actions (`seq`), nondeterministic updates (`choose`).

Functions which are not updated by rule transitions are *static*. Those updated are *dynamic*, and distinguished in *monitored* (read by the machine and modified by the environment), *controlled* (read and written by the machine).

ASMs allow modeling different computational paradigms, from a *single* agent executing parallel actions, to distributed *multiple* agents interacting in a synchronous or asynchronous way. We exploit the latter computational model for modeling security protocols. More specifically, a *multi-agent ASM* is a family of pairs $(a, ASM(a))$, where each agent $a : Agent$ executes its own machine $ASM(a)$ specifying its local behavior, and contributes to determine the next state. A predefined function *program* on *Agent* indicates the ASM associated with an agent. As an example, Fig. 2 shows an excerpt of the multi-agent ASM models of the two case studies, where the three agents, `agentA` (for principal A), `agentB` (for principar B) and the attacker `agentE`, operate in a distributed setting according to the protocol rules (see Sect. 2.1 for more details).

Within transition rules, each agent can identify itself by means of a special 0-ary function *self*: *Agent* which is interpreted by each agent a as itself. An ASM agent can behave according to a *control-state ASM* [14]: transition rules are guarded by a function *mode*, which is updated in the rule body and whose values resemble the states of a Finite State Machine. This machine model has been used for specifying principal behaviour (i.e., actor's actions).

ASMETA Toolset. The ASM formal method is supported by the tool-set ASMETA (ASM mETAmodeling) [4] for model editing, validation and verification.

An ASM model edited in ASMETA by means of the AsmetaL notation [21], has a predefined structure consisting of: a *signature*, which contains declarations of domains and functions; a block of *definitions* of static domains and functions, transition rules, state invariants and properties to verify; a *main rule*, which is the starting point of a machine computation; a set of *initial states*, one of which is elected as *default* and defines an initial value for the machine locations.

An AsmetaL model can include other pieces of ASM specifications imported as module from the main machine (that is declared as *asm*). Every module contains definitions of domains, functions, invariants and rules, while the main AsmetaL model is a module that additionally contains an initial state and a main rule representing the starting point of the execution. The module importing mechanism allows for the specification of predefined *libraries* of signature declaration and definition, as those provided in our approach as predefined set of modeling primitives for cryptography protocols specification.

Among the tools of the ASMETA framework, the model checker AsmetaSMV can be used to check if given *properties*, expressed within the model as temporal logic formulas, hold during all possible model executions. Here, we use AsmetaSMV to verify the CIA properties for the two case studies (see Sect. 3.3). In case a property does not hold, a counterexample is returned by the model checker, and thanks to integration of the ASMETA tools, counterexamples obtained from false properties are given back as model *scenarios* that are simulated (an example is shown in Sect. 3.3). Besides the greater expressiveness of the ASM mathematical notation w.r.t. the limited modeling primitives of a model checker, the advantage of using ASM/ASMETA for our purposes, instead of directly specifying protocols in the input language of the model checker (e.g., NuSMV), is strictly related with the tool integration feature of ASMETA.

3   ASM Modeling of Cryptographic Protocols

We here introduce the main primitives (in terms of ASM functions and domains in the AsmetaL notation) useful to model cryptographic protocol. These primitives can be imported as a library, called CryptoLibrary, in the AsmetaL model of a specific protocol; library domains can be instantiated for the specific purposes of the protocol, while functions can be directly used within the model.

To explain library primitives and show their use for modeling the two case studies, we use the following convention: excerpts are reported in listings; they are colored in gray (and not numbered) if they refer to signature declared in the library, and are numbered and uncolored if excerpts refer to a model.

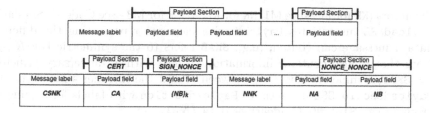

Fig. 1. Structure of a message and examples in SSL and NSPK protocols

Actors or Principals. The library specifies protocol actors as `agentX` of corresponding domain `Agent_X`, where `X` identifies the role of the agent working in the protocol as A (`X=A`), as B (`X=B`), as a malicious actor (`X=E`), and as a trusted key distribution server (`X=S`).

```
domain Agent_A subsetof Agent
domain Agent_B subsetof Agent
domain Agent_E subsetof Agent
domain Agent_S subsetof Agent

static agentA: Agent_A
static agentB: Agent_B
static agentE: Agent_E
static agentS: Agent_S
```

Messages. Protocol messages are modeled by means of library domains and functions, according to the message *structure* given in Fig. 1 (the picture above shows the general structure, the two pictures below exemplify the structure of SSL message M3, on the left, and of NSPK M2, on the right). The domain `Message` is used to model message labels; `Payload` is the set of protocol payload data (e.g., nonces, certificates, keys, etc.); the domain `PayloadSection` contains labels indicating groups[1] of payload data; the function `protocolMessage` associates to a (*sender*, *receiver*) pair of actors the *label* of the message they are exchanging; the function `payload` yields the set of *payload data* contained in the *payload section* of an exchanged message of a given *label*.

Library domains `Message`, `Payload` and `PayloadSection` will be instantiated according to the characteristics of the specific protocol to model[2].

```
domain Message subsetof Any
domain Payload subsetof Any
domain PayloadSection subsetof Any
controlled protocolMessage: Prod(Agent,Agent)—> Message
controlled payload: Prod(Message,PayloadSection, Agent, Agent) —> Powerset(Payload)
```

For example, in case of the SSL protocol, `Message` domain is instantiated as `MessageSSL`[3] containing the labels of the three messages exchanged between

---

[1] A *group* contains either data subject to a specific protocol operation (e.g., signature, hashing, etc.), or data with no further manipulation.

[2] *Any* stands for a domain that can contain any element.

[3] We convey to add the suffix naming the protocol to the corresponding library domain when we instantiate it.

the actors (KK for message M1, NK for M2, CSNK for M3 - see Code 1). Similarly, PayloadSSL instantiates Payload with elements representing the different data a message can contain (e.g., SKAB refers to the symmetric key K_{AB}). Note that it also contains information to deal with an adversary principal (as SKAE, symmetric key K_{AE} between A and the intruder - see Sect. 3.1). PayloadSectionSSL instantiates PayloadSection with labels for significant groups (e.g., SIGN_NONCE identifies data $\{N_B\}_{K_A^{-1}}$).

```
enum domain MessageSSL = {KK | NK | CSNK}
enum domain PayloadSSL = {NB|CA|SKAB|SKAE|SKEB}
enum domain PayloadSectionSSL={SEND_KEY|NONCE|CERT|SIGN_NONCE}
....
protocolMessage(agentA,agentB):= KK
payload(KK,SEND_KEY,agentA,agentB):=SKAB
```

Code 1. Excerpt of ASM model for SSL

Actor's Actions. Each (honest) protocol actor runs a program having the same structure; this reflects the common pattern that crypto-protocols have. Figure 2 in Sect. 3.2 shows the pattern similarity in the two case studies, SSL and NSPK protocols. Each rule in the actor's program (e.g., r_agentARule for actor agentA) models the role that the actor has in the protocol run. Each actor has a *memory*: (s)he keeps information related to the protocol session by means of the following built-in functions available in the library.

```
controlled knownAsimPubKey: Agent −> Powerset(AsimPubKeyType)
controlled knownAsimPrivKey: Agent −> Powerset(AsimPrivKeyType)
controlled knownSimmKey: Agent −> Powerset(SimmKeyType)
controlled knownPayload: Agent −> Powerset(PayloadType)
```

During a protocol session, an actor follows a control-state ASM on the base of the mode functions internalState on State and protocolMessage on Message: each control state represents the actor's configuration upon performing a protocol step where (s)he plays a specific role (as A or B).

```
domain State subsetof Any
controlled internalState: Agent −> State
```

Primitives internalState and State must be instantiated to model a specific protocol. The excerpt in Code 2 shows the instantiation for the SSL protocol. It also shows a rule of the (agentB's) control-state that causes the actor's state change. The other rules have similar guarded structures.

```
enum domain StateA = {IDLE_A | WAITING_NK | END_A}
enum domain StateB = {WAITING_KAB | WAITING_CSNK | END_B}
controlled internalStateA: Agent_A −> StateA
controlled internalStateB: Agent_B −> StateB

if(internalStateB(self)= WAITING_KAB and protocolMessage(e,self)= KK)then
    par
        protocolMessage(self,e):= NK
        payload(NK,NONCE,self,e):= {NB}
        internalStateB(self):= WAITING_CSNK
    endpar
endif
```

Code 2. Control state ASM model of an Agent in SSL

Symmetric/Asymmetric encryption primitives. Built-in primitives exist for handling encryption/decryption of messages by using symmetric and asymmetric (public and private) keys.

```
domain SimmKey subsetof Any
controlled encrypt_sim: Prod(Message,Powerset(PayloadSection),Agent,Agent) −> SimmKey
static decrypt_sim: Prod(Message,Powerset(PayloadSection), Agent, Agent) −> Boolean
static sim_keyAssociation: Prod(Agent,Agent) −> SimmKey
    function decrypt_sim(m in Message, s in Powerset(PayloadSection), a1 in Agent, a2 in Agent)=
    if (contains(knownSimmKey(a2),encrypt_sim(m,s,a1,a2 ))) then true else false endif
```

```
domain AsimPubKey subsetof Any
domain AsimPrivKey subsetof Any
controlled encrypt_asim: Prod(Message,Powerset(PayloadSection),Agent,Agent) −> AsimPubKey
static asim_revKeyAssociation: AsimPrivKey −>AsimPubKey
static asim_keyAssociation: AsimPubKey −> AsimPrivKey
static decrypt_asim: Prod(Message,Powerset(PayloadSection),Agent,Agent) −> Boolean
    function decrypt_asim(m in MessageType,s in Powerset(PayloadSectionType),a1 in Agent,
                         a2 in Agent)=
    if (contains(knownAsimPrivKey(a2),asim_keyAssociation(encrypt_asim(m,s,a1,a2)))) then
       true else false endif
```

In both symmetric and asymmetric cases, function `encrypt_[a]sim` memorises the *encryption key* used to encrypt a message of a given *label* and *payload*, sent by a *sender* actor to a *receiver* one. For example, in case of protocol SSL, when message M3. $A \to B : \{C_A, \{N_B\}_{K_A^{-1}}\}_{K_{AB}}$ is sent, location `encrypt_sim(CSNK,{CERT,SIGN_NONCE},agentA,agentB)` is updated with SKAB, being CSNK the message label, `{CERT,SIGN_NONCE}` the message payload, `agentA` the sender, `agentB` the receiver, and SKAB the two agents symmetric encryption key.

Decryption functions are predicates used within rule guards to check if an actor receiving an encrypted message of a given *label* and *payload section* can decrypt the received message: it is *false* when the message encryption key is not contained in the agent's memory.

The following Code 3 shows the rule `r_message_NK` of the SSL model, which is executed by actor B at step M2 of the protocol under the Dolev-Yao assumption (in this case, the traffic is controlled by intruder E, therefore, B views all messages as sent by E - see Sect. 3.1); function `decrypt_sim` is used to check if B can decrypt the message, and `encrypt_sim` yields the symmetric key used for encryption.

```
rule r_message_NK =
   let (e=agentE) in
      if (internalStateB(self)=WAITING_KAB and protocolMessage(e,self)=KK) then
         if decrypt_asim(KK,{SEND_KEY},e,self) then
         par
            knownSimmKey(self):= union(knownSimmKey(self),payload(KK,SEND_KEY,e,self))
            protocolMessage(self,e):= NK
            payload(NK,NONCE,self,e):= {NB}
            encrypt_sim(NK,{NONCE},self,e):= SKAB
            internalStateB(self):= WAITING_CSNK
         endpar
         endif endif endlet
```

Code 3. ASM rule of an Agent in SSL

Signature. In case of asymmetric encryption, the function `sign` is used to memorize the *signature* of a message of given *label* and *payload section*, sent by a *sender* actor to a *receiver* actor; predicate `verify_sign` is used to check if an actor receiving a message containing signed data can verify the authenticity of the message: it checks if the actor verifying the signature knows the sender's public key and the signed data matches the content of the signature.

```
controlled sign: Prod(MessageType, PayloadSectionType, Agent, Agent) −> AsimPrivKeyType
static verify_sign: Prod(MessageType,PayloadSectionType, Agent, Agent) −> Boolean
function verify_sign(m in Message, s in PayloadSection, a1 in Agent,a2 in Agent)=
    if (contains(knownAsimPubKey(a2),asim_revKeyAssociation(sign(m,s,a1,a2)))) then
        if (allin(knownPayload(a2),payload(m,s,a1 ,a2))) then true else false endif
    endif
```

Advanced features. The library makes also available functions to handle advanced characteristics of crypto-protocols, e.g., hash functions, MAC functions, the Diffie-Hellman (to name a few); these have not been used in the two case studies reported here and, therefore, they are not presented here.

3.1   Modeling the Intruder Behaviour

We support two possible behaviours of an *intruder*: *a)* as a *protocol actor* having malicious behaviour, and *b)* as an *external actor* operating in *passive* mode (i.e., an eavesdropper), or in *active* mode according to the Dolev-Yao assumption. The intruder is available in the library as `agentE`. His/her program reflects the intended behaviours. Note that, the attacker controls the traffic and this is modeled by imaging E in the middle of the communication between A and B. Therefore, in the presence of the attacker (internal or external), the channel between the two honest actors is broken in two sub-channels: between A and E, and between E and B (this justifies why, in the rules, all messages sent by A are received by E, and all messages received by B arrive from E).

Protocol actor with a malicious behaviour. In this case, `agentE`, operating as an intended protocol actor, receives messages from A who voluntarily started a protocol session with E. In this case, (s)he can save the information of the payload, but instead of honestly replying to A following the protocol rules, E uses his/her information to start a new protocol session with B. The library monitored function `chosenReceiver` can be used to capture both the case in which the intended receiver behaves honestly (as B), and the case in which (s)he behaves dishonestly (as E). Domain `Receiver` contains those agents that can be selected as possible protocol receivers.

```
enum domain Receiver = {agentB|agentE}
monitored chosenReceiver: Receiver
controlled receiver: Receiver
```

External attacker. In this case, `agentE` is an external actor, thus A cannot start a protocol session with E. Information known by the intruder come from what (s)he is able to steal from the traffic or from a prior knowledge. The attacker can operate in `PASSIVE` or `ACTIVE` mode. This operation `mode` can be selected by the monitored function `chosenMode` and the same protocol can be analysed in the two passive/active intruder scenarios.

```
enum domain Modality = {ACTIVE | PASSIVE}
monitored chosenMode: Modality
controlled mode: Modality
```

In PASSIVE mode, `agentE` can view all the messages of the communication between `agentA` and `agentB`, save all information transmitted in clear and those contained in message payloads that (s)he is able to decrypt. In ACTIVE mode, `agentE` operates under the Dolev-Yao assumption and controls the traffic: besides the capabilities as passive attacker, (s)he can craft messages by using information stolen (from messages sent by `agentA` or `agentB`) and send fake messages (consistent with the protocol rules).

For each message type, the intruder has rules for working in *passive* or *active* mode. Rules `r_message_eavesdrop` model the intruder operating as passive attacker; rules `r_message_craft` model the intruder operating as external active actor and as protocol actor with malicious behavior (in both cases, (s)he is able to craft new messages).

Unlike `agentA` and `agentB`, `agentE` has no `internalState` and does not work according to a control-state ASM. His/her behaviour depends on what (s)he views in the traffic. As the other agents, `agentE` has a knowledge (`knownSimmKey`, `knownAsimPrivKey`, `knownAsimPubKey`, `knownPayload`) that can be exploited to steal information or craft fake messages. To endow the intruder with a prior knowledge, it is enough to set initial values for the knowledge functions.

```
rule r_agentARule =                    rule r_agentARule =
   par                                    par
      r_message_KK[]                          r_message_NAK[]
      r_message_CSNK[]                        r_message_NK[]
   endpar                                 endpar
rule r_agentBRule =                    rule r_agentBRule =
   par                                    par
      r_message_NK[]                          r_message_NNK[]
      r_check_CSNK[]                          r_check_NK[]
   endpar                                 endpar
rule r_agentERule =                    rule r_agentERule =
   par                                    par
      r_message_eavesdrop_KK[]                r_message_eavesdrop_NAK[]
      r_message_eavesdrop_NK[]                r_message_eavesdrop_NNK[]
      r_message_eavesdrop_CSNK[]              r_message_eavesdrop_NK[]
      r_message_craft_KK[]                    r_message_craft_NAK[]
      r_message_craft_NK[]                    r_message_craft_NNK[]
      r_message_craft_CSNK[]                  r_message_craft_NK[]
   endpar                                 endpar
main rule r_Main =                     main rule r_Main =
   par                                    par
      program(agentA)                         program(agentA)
      program(agentB)                         program(agentB)
      program(agentE)                         program(agentE)
   endpar                                 endpar
```

Fig. 2. ASM models of protocols SSL and NSPK

3.2 NSPK and SSL Models

Figure 2 reports the high level ASM models of the two case studies by using predefined domains and functions of the library `CryptoLibrary` and instantiating domains when required. The two models have the same structure: each one is a multi-agent ASM with programs for honest actors `agentA` and `agentB` – according to the actor's role in the protocol–, and for the attacker `agentE`; each honest actor has a rule for each kind of protocol message (s)he has to build, plus the final *B*'s rule for the final message check; the attacker has rules to handle all possible behaviours (as eavesdropper and as message crafter) as described before. Call rules definitions differ depending to the protocol.

Code 4 shows, as an example, an excerpt of the rule `r_message_CSNK` of protocol SSL, in the case of *A* talking with *B*. The rule is executed by `agentA` when she is in (internal) state `WAITING_NK` and receives a message of label `NK` (step M2): she can decrypt the message and sends a message of label `CSNK`, which is encrypted by the symmetric key between herself and the expected receiver `agentB`; the message has two payload sections: one of label `CERT`, which has payload `CA`, and one of label `SIGN_NONCE`, which has the same payload (i.e., NB) of the previously sent message having label `NK` and payload section `NONCE`, signed by the private actor's key `PRIVKA`.

```
rule r_message_CSNK=
let (e=agentE) in
    if (internalStateA(self)=WAITING_NK and protocolMessage(e,self)=NK) then
        if (receiver=agentB) then
            if (decrypt_sim(NK,{NONCE},e,self)=true) then
                par
                    protocolMessage(self,e):= CSNK
                    payload(CSNK,CERT,self,e):= {CA}
                    payload(CSNK,SIGN_NONCE,self,e ):= payload(NK,NONCE,e,self)
                    sign(CSNK,SIGN_NONCE,self,e):= PRIVKA
                    encrypt_sim(CSNK,{CERT,SIGN_NONCE},self,e):= sim_keyAssociation(self,receiver)
                    internalStateA(self):= END_A
                endpar
            endif
        endif
..... endif endlet
```

Code 4. An excerpt of a rule of actor `agentA` in SSL protocol

3.3 Security Properties Schema

In this section, we introduce schema also for specifying in CTL formulas the security goals of a protocol in terms of the CIA triad.

Confidentiality. In communication protocols, encryption is used to obtain data confidentiality: only actors with correct decryption keys are able to access data.

Let *x* be a given exchanged data, *confidentiality* of *x* is assured by proving that *there is not a state in the future in which E knows x*:

$$\neg EF(\ \texttt{contains(knownPayload(agentE), x)})$$

Therefore, the confidentiality of the nonce N_B in the protocol NSPK is:

$$\neg EF(\text{ contains(knownPayload(agentE), NB)})$$

This property is false in the NSPK protocol. Code 5 shows a scenario of the NSPK protocol model, which reflects the counterexample reported by the model checker: if `agentA` initiates a protocol run with `agentB` under the Dolev-Yao assumption of an active attacker `agentE` (the `set` command is used to assign specific values to monitored functions) controlling the traffic, upon a certain number (not reported here) of machine steps (the command `step` forces a step of the ASM execution), the protocol run ends with both agents `agentA` and `agentB` successfully completing the protocol, but also with the shared secret NB known by the attacker (the `check` command is used to inspect location values in the current state of the underlying ASM).

Note that if message M2 is patched using message $\{N_A, N_B, B\}_{K_A}$, the property holds.

```
scenario NB_interception
load NS/NS.asm

set chosenReceiver:=agentE;
set chosenMode:=ACTIVE;
step
check internalStateA(agentA)=IDLE_A and internalStateB(agentB)=WAITING_NAK;
step
check internalStateA(agentA)=WAITING_NNK and
internalStateB(agentB)=WAITING_NAK and protocolMessage(agentA,agentE)= NAK
and payload(NAK,NONCE_ID,agentA,agentE)={NA,ID_A} and
encrypt_asim(NAK,{NONCE_ID},agentA,agentE)=PUBKE;
step
.....
step
check internalStateA(agentA)=END_A and internalStateB(agentB)=END_B;
check contains(knownPayload(agentE),NB);
```

Code 5. Scenario from the counterexample of the false confidentiality property

Integrity. Exchanged message has not been altered. Thus, a message can be considered integral when the payload sent is the same as the payload received.

Let m be a message exchanged. Integrity of m is assured by proving that *there is not a state in the future in which E altered the payload section sec of the message m:*

$$\neg EF(\text{ payload(m,sec,agentA,agentE)=payload(m,sec,agentE,agentB)})$$

In none of the two protocols there is integrity violation, since there are two separate sessions and the attacker does not modify any message along the channel.

Authentication. A principal proves his/her identity by demonstrating knowledge of a secret (not necessarily shared) and of an information that varies over time (to avoid replay attacks), without explicitly revealing the secret.

Let x be a secret that must be shared by the two actors at the end of the protocol execution. Authentication of entity x is assured by proving that *if there is a state in the future in which A and B know x, then E does not know x.*

$$EF(\text{ contains(knownPayload(agentB), x) and contains(knownPayload(agentA), x))} \rightarrow$$
$$AG(\text{ }\neg(\text{contains(knownPayload(agentE), x)}))$$

The authentication of the secret nonce N_B in the protocol SSL is:

$$\boldsymbol{EF}(\;\texttt{contains(knownPayload(agentB), NB)}\text{ and }\texttt{contains(knownPayload(agentA), NB)})\longrightarrow$$
$$\boldsymbol{AG}(\;\neg(\texttt{contains(knownPayload(agentE), NB)}))$$

This property does not hold; if message M3 is patched using the message $\{C_A, \{A, B, K_{AB}, N_B\}_{K_A^{-1}}\}_{K_{AB}}$, the property is true.

4 Related Work

After the seminal paper on BAN logic [15] introducing one of the first formalisms designed to reason about protocols, many techniques and tools have been proposed for verifying protocols.

TAMARIN [26] is a well-known tool dedicated to the formal analysis of security protocols. It has been used to verify complex protocols such as TLS, 5G, or RFID protocols. It uses its own modelling language to specify protocols, adversary models and security properties. For the verification process, it is based on deduction and equational reasoning. However, one of its drawback is its lack of full automation: for many protocols, the user needs to write intermediate lemmas (called sources lemmas), or to manually guide the proof search. In [16], the authors propose a technique to automatically generate sources lemmas that works for simple protocols, but still needs user interaction in case of large protocols. Scyther [17] is somehow an extension of TAMARIN, with a more accessible and explicit modelling language by means of a graphical user interface.

ProVerif [11] uses applied pi-calculus [27] as modelling language, and a resolution algorithm on Horn clauses for the verification. It may need human intervention when a protocol proof fails because of some internal approximations, or it may find false attacks. However, it can verify protocols without arbitrarily bounding the number of executed protocol sessions or the message size.

Verifpal [24] is a more recent framework inspired by ProVerif. As us, the authors recognise that the limited usage of formal methods is mainly due to the fact that the languages employed by most tools is too difficult and abstract. Verifpal uses a simple modelling language similar to the usual protocol notation that we introduced in Sect. 2.1 at the cost of some compromises in analysis completeness. For more sound proofs, translation of models in ProVerif and Coq [10] are possible. The tool's internal logic relies on the deconstruction and reconstruction of abstract terms, similar to existing symbolic verification tools.

AVISPA (Automated Validation of Internet Security Protocols and Applications) [5] is a platform that groups several tools, using different verification techniques. It uses HLPSL (High-Level Protocol Specification Language) as protocol description language. The HLPSL model is then translated in an intermediate model, that is given as input to four tools: SATMC (SAT-based Model-Checker) [6] for a bounded state space, CL-AtSe and OFMC for a bounded number of sessions, TA4SP for an unbounded number of sessions.

The use of ASMs to model security protocols has been tackled in the past. However, most of the works are rather outdated and they consist on a single case study, rather than a general approach, in many cases with a limited tool support. In [7], a model of the Kerberos protocol is given through stepwise refinements of ASMs. In [22] an interactive theorem prover is used to prove security properties on a smart card application. Although no proof of standard CIA security properties is given, the authors recognise, as we do, how the ASM refinement theory could be used to translate the protocol ASM model into a (security-proven) Java implementation. Our closest ASM-based approach is [1], where an attempt to generalise the attacker model is given. However the verification technique is by means of invariant checking via simulation-based attack scenarios.

The idea of patterns reusing is not new as well. In [9], a systematic way to design secure-by-construction cryptographic protocols, where the proof process reuses smaller protocol parts previously proven to be correct and secure. In this case the approach is based on the B notation. The work of [8] describes a method for implementing and analyse a specific class of security protocols (i.e., classical key distribution protocols) in SPIN. In particular, the authors focus on modelling a generic intruder model working with all the protocols within the class. The work in [28] presents a model-driven approach to design security-critical systems based on cryptographic protocols and to prove application-specific security properties. A smart card application is analyzed. UML is used as front-end modeling notation, whereas ASMs and a theorem prover are used as back-end formalisms for property verification; the underlying idea is similar to our.

5 Conclusion

In this paper we presented an ASM-based approach for the automatic verification of cryptographic protocols. To verify the feasibility of our approach, we tested it both with small protocols and more difficult ones, for example the Z-Wave protocol in the IoT scenario [25] (where we found a vulnerability that has been confirmed by the Z-Wave Alliance). With respect to other approaches based on deduction and equational reasoning, at present, we are able to verify protocols with a bounded number of sessions. In addition, as for all tools based on model-checking, we are able to find attacks against protocols, but not to prove the absence of attacks, since attacks may appear in an unexplored part of the state space. However, in our case no human intervention is requested to help the tool to end the verification session, and we never get false positives.

In order to make the approach more user-friendly, the next step is to build a GUI connecting each element defined in the library with a graphical counterpart (for example expressed as a sequence diagram). We would also like to translate our models in prototype Java implementations by exploiting ASMETA toolset.

References

1. Al-Shareefi, F.: Analysing safety-critical systems and security protocols with abstract state machines. Ph.D. thesis, University of Liverpool (2019)

2. Arcaini, P., Gargantini, A., Riccobene, E.: AsmetaSMV: a way to link high-level ASM models to low-level NuSMV specifications. In: Frappier, M., Glässer, U., Khurshid, S., Laleau, R., Reeves, S. (eds.) ABZ 2010. LNCS, vol. 5977, pp. 61–74. Springer, Heidelberg (2010). https://doi.org/10.1007/978-3-642-11811-1_6

3. Arcaini, P., Gargantini, A., Riccobene, E.: Rigorous development process of a safety-critical system: from ASM models to Java code. Int. J. Softw. Tools Technol. Transf. **19**(2), 247–269 (2017)

4. Arcaini, P., Gargantini, A., Riccobene, E., Scandurra, P.: A model-driven process for engineering a toolset for a formal method. Softw. Pract. Exp. **41**(2), 155–166 (2011)

5. Armando, A., et al.: The AVISPA tool for the automated validation of internet security protocols and applications. In: Etessami, K., Rajamani, S.K. (eds.) CAV 2005. LNCS, vol. 3576, pp. 281–285. Springer, Heidelberg (2005). https://doi.org/10.1007/11513988_27

6. Armando, A., Compagna, L., Ganty, P.: SAT-based model-checking of security protocols using planning graph analysis. In: Araki, K., Gnesi, S., Mandrioli, D. (eds.) FME 2003. LNCS, vol. 2805, pp. 875–893. Springer, Heidelberg (2003). https://doi.org/10.1007/978-3-540-45236-2_47

7. Bella, G., Riccobene, E.: Formal analysis of the Kerberos authentication system. J. Univ. Comput. Sci. **3**(12), 1337–1381 (1997)

8. Ben Henda, N.: Generic and efficient attacker models in SPIN. In: Proceedings of International SPIN Symposium on Model Checking of Software, pp. 77–86 (2014)

9. Benaissa, N., Méry, D.: Cryptographic protocols analysis in event B. In: Pnueli, A., Virbitskaite, I., Voronkov, A. (eds.) PSI 2009. LNCS, vol. 5947, pp. 282–293. Springer, Heidelberg (2010). https://doi.org/10.1007/978-3-642-11486-1_24

10. Bertot, Y., Castran, P.: Interactive Theorem Proving and Program Development: Coq'Art The Calculus of Inductive Constructions, 1st edn. Springer, Heidelberg (2010)

11. Blanchet, B.: An efficient cryptographic protocol verifier based on prolog rules. In: Proceedings of IEEE Computer Security Foundations Workshop, pp. 82–96 (2001)

12. Bonfanti, S., Gargantini, A., Mashkoor, A.: Design and validation of a C++ code generator from abstract state machines specifications. J. Softw. Evol. Process. **32**(2) (2020)

13. Börger, E., Raschke, A.: Modeling Companion for Software Practitioners. Springer, Heidelberg (2018). https://doi.org/10.1007/978-3-662-56641-1

14. Börger, E., Stärk, R.: Abstract State Machines: A Method for High-Level System Design and Analysis. Springer, Heidelberg (2003). https://doi.org/10.1007/978-3-642-18216-7

15. Burrows, M., Abadi, M., Needham, R.: A logic of authentication. ACM Trans. Comput. Syst. **8**(1), 18–36 (1990)

16. Cortier, V., Delaune, S., Dreier, J.: Automatic generation of sources lemmas in TAMARIN: towards automatic proofs of security protocols. In: Chen, L., Li, N., Liang, K., Schneider, S. (eds.) ESORICS 2020. LNCS, vol. 12309, pp. 3–22. Springer, Cham (2020). https://doi.org/10.1007/978-3-030-59013-0_1

17. Cremers, C.J.F.: The scyther tool: verification, falsification, and analysis of security protocols. In: Gupta, A., Malik, S. (eds.) CAV 2008. LNCS, vol. 5123, pp. 414–418. Springer, Heidelberg (2008). https://doi.org/10.1007/978-3-540-70545-1_38

18. Davis, J.A., et al.: Study on the barriers to the industrial adoption of formal methods. In: Pecheur, C., Dierkes, M. (eds.) FMICS 2013. LNCS, vol. 8187, pp. 63–77. Springer, Heidelberg (2013). https://doi.org/10.1007/978-3-642-41010-9_5

19. Dolev, D., Yao, A.: On the security of public key protocols. IEEE Trans. Inf. Theory **29**(2), 198–208 (1983)
20. Domo: Data never sleeps 6th (2017)
21. Gargantini, A., Riccobene, E., Scandurra, P.: A metamodel-based language and a simulation engine for abstract state machines. J. UCS **14**(12), 1949–1983 (2008)
22. Haneberg, D., Grandy, H., Reif, W., Schellhorn, G.: Verifying security protocols: an ASM approach. In: Proceedings of International Workshop on Abstract State Machines (2005)
23. Haskins, B., Stecklein, J., Dick, B., Moroney, G., Lovell, R., Dabney, J.: 8.4.2 error cost escalation through the project life cycle. In: INCOSE International Symposium, vol. 14, pp. 1723–1737 (2004)
24. Kobeissi, N., Nicolas, G., Tiwari, M.: Verifpal: cryptographic protocol analysis for the real world. In: Bhargavan, K., Oswald, E., Prabhakaran, M. (eds.) INDOCRYPT 2020. LNCS, vol. 12578, pp. 151–202. Springer, Cham (2020). https://doi.org/10.1007/978-3-030-65277-7_8
25. Lilli, M.: Formal verification of Z-Wave protocol security properties. Master's thesis, Università degli Studi di Milano, Italy (2020)
26. Meier, S., Schmidt, B., Cremers, C., Basin, D.: The TAMARIN prover for the symbolic analysis of security protocols. In: Sharygina, N., Veith, H. (eds.) CAV 2013. LNCS, vol. 8044, pp. 696–701. Springer, Heidelberg (2013). https://doi.org/10.1007/978-3-642-39799-8_48
27. Milner, R.: Communicating and Mobile Systems: The π-calculus. Cambridge University Press, Cambridge (1999)
28. Moebius, N., Stenzel, K., Reif, W.: Generating formal specifications for security-critical applications - a model-driven approach. In: Workshop on Software Engineering for Secure Systems, pp. 68–74 (2009)

Verifying System-Level Security
of a Smart Ballot Box

Dana Dghaym[(✉)], Thai Son Hoang, Michael Butler, Runshan Hu,
Leonardo Aniello, and Vladimiro Sassone

ECS, University of Southampton, Southampton, UK
{d.dghaym,t.s.hoang,m.j.butler,rs.hu,l.aniello,vsassone}@soton.ac.uk

Abstract. Event-B, a refinement-based formal modelling language, has
traditionally focused on safety, but now increasingly finds a new role in
developing secure systems. In this paper we take a fresh look at security
and focus on what security means for the system rather than looking at
detailed protocols. We use Event-B for proving security from an abstract
view and refining it towards design details, focusing on the refinement of
the availability property of the system. We define a general approach to
guarantee the availability of events by ensuring the *non-strengthening* of
their guards, taking into consideration their parameter types. We illustrate our approach using a smart ballot system, an integral part of modern voting systems.

Keywords: Event-B · Availability property · System security ·
Refinement · Voting system

1 Introduction

Event-B [1] is a refinement-based formal method for developing discrete transition systems. Data refinement is a standard technique in Event-B that requires
relating the abstract variables with the concrete variables using gluing invariants. To ensure the refinement correctness, a guard strengthening (GRD) Proof
Obligation (PO) is generated to verify that if a concrete event is enabled then
its corresponding abstract event is also enabled.

In this paper, we investigate the application of refinement-based formal modelling in building a correct-by-construction secure system. Our focus is the refinement of the *availability property* of secure systems. The availability property of
a refined event can be proved, if conditions under which event is enabled in an
abstract machine are maintained in the refined machine.

We illustrate our approach using a smart ballot box case study. In this case
study we build a secure system by gradually introducing the confidentiality and
integrity properties of the voting system using encryption and message authentication respectively, while ensuring availability throughout the refinement process. In the smart ballot box system case study, confidentiality is ensured by
having the voter's choices not visible to the system, while integrity is ensured by

A. Raschke and D. Méry (Eds.): ABZ 2021, LNCS 12709, pp. 34–49, 2021.
https://doi.org/10.1007/978-3-030-77543-8_3

only accepting valid ballots and only rejecting invalid ballots. The availability of the system is guaranteed by not preventing a voter from casting a valid ballot. In this case study, we also model the intruder's behaviour and show how encryption and authentication can provide protection against an intruder behaviour.

We propose an extension to Event-B Proof Obligations (POs) that relates the enabledness of an event to availability of behaviour according to a new parameter type, which we call a *rigid* parameter. This PO will ensure the availability property of events is preserved by refinement.

The rest of the paper is structured as follows. Section 2 gives an overview of the smart ballot box case study. Section 3 introduces Event-B formal method. We propose a new PO to prove the availability of refined events in Sect. 4. In Sect. 5, we present our Event-B development of the case study across the different refinements. We discuss how model checking improved our development in Sect. 6. We compare our approach with other work in Sect. 7. Finally, we conclude and present our plans for future work in Sect. 8.

2   Case Study: Smart Ballot Box

The main function of the Smart Ballot Box (SBB) [5] is to inspect a ballot paper by detecting a 2D barcode, decode it and evaluate if the decoded contents verifies the paper from a Ballot Marking Device (BMD). If the ballot is valid, then it can be cast into the storage box. Otherwise, the SBB rejects the paper, that will be ejected. The SBB does not conduct a full-scale analysis of the document, nor record the choices of the voters, nor tabulate the votes of the ballots it scans. The key function of the SBB is to ensure that only valid countable summary ballot documents that can be tabulated later are included in ballot boxes.

REQ 1 The BMD is used by a voter to make ballot choices and print a summary ballot. The ballot choices are not recorded by the BMD.
REQ 2 The barcode is created by the BMD as an authenticator so that the SBB can recognise a legitimate ballot.
REQ 3 The ballot barcode is formed of a timestamp followed by an encoding of the encrypted ballot and the Message Authentication Code (MAC).
REQ 4 All ballots with invalid or non-existent barcode are rejected by the SBB.
REQ 5 All ballots with a valid barcode are recognized by the SBB.
REQ 6 The user can decide whether to cast or spoil the given ballot. The ballot is deposited into the ballot box if the voter decides to cast their vote and subsequently all ballots with the same barcode will be considered invalid.
REQ 7 The user can decide to spoil a valid ballot. In such cases, the ballot is ejected and returned to the voter, and is subsequently considered invalid.
REQ 8 The ballot box shall reject a ballot with an expired barcode.
REQ 9 The authentication scheme is based on a MAC created with the AES standard encryption algorithm and a cryptographic key shared by the BMD and SBB. The SBB authenticates the ballot by recreating the MAC with the shared key and comparing the result with the MAC encoded in the barcode. If the two MAC values are equal the ballot is considered valid.

3   Background

Event-B [1] is a refinement-based formal method for system development. An Event-B model contains two parts: *contexts* for static data, and *machines* for dynamic behaviour specified by *variables* v, *invariant* predicates I(v) that constrain the variables, and *events*. An event comprises a guard denoting its enabling condition and an action describing how the variables are modified when the event is executed. In general, an event e has the following form, where t are the event parameters, $G(t, v)$ is the event guard, and $v := E(t, v)$ is the action of the event.

$$\textbf{any t where } G(t,v) \textbf{ then } v := E(t,v) \textbf{ end}$$

Refinement in Event-B is reasoned event-wise where the behaviour of the concrete or refining machine conforms with the abstract machine. This is ensured by the refinement rules of guard strengthening and action simulation of the abstract events by their corresponding refining events. In addition to gluing invariants that relate the abstract and refined state of the models. In this case, given the abstract invariant I(v) and gluing invariant J(v,w) (where v and w are abstract and concrete variables, respectively), the GRD PO will check that the guards of the concrete event H(q,v,w) are stronger than the guards of its corresponding abstract event G(p,v) (where p and q are the abstract and concrete parameters, respectively) as follows:

$$I(v), J(v,w), H(q, v, w) \vdash \exists p \cdot G(p,v) \ ,$$

The GRD PO will ensure that if a concrete event is enabled, then its corresponding abstract event must be enabled. In Event-B, an event is enabled if there are parameter values that satisfy the guard of the event.

Event-B is supported by the Rodin Platform (Rodin) [2], an extensible toolkit which includes facilities for modelling, verifying the consistency of models using theorem proving and model checking techniques, and validating models with simulation-based approaches. In this paper we use both Event-B theorem proving and model checking using ProB [9] for the validation and verification of the SBB.

4   Rigid Events and Parameters

In this section, we first introduce the notion of enabledness with respect to a set of parameters (Sect. 4.1). Subsequently, we elaborate on the meaning of availability properties and introduce the notion of rigid events and parameters to capture availability properties (Sect. 4.2). We then address preservation of availability properties during refinement in Sect. 4.3.

4.1   Event Enabledness and Parameters

We extend the notion of event enabledness in Event-B to include the parameters. Given an event e with parameters p and q (both p and q can be a list of variables) and G(p, q), the guard of e constraining p and q. We define Enabled$_p$ (e) as follows.

$$\text{Enabled}_p(e) \mathrel{==} \exists q \cdot G(p, q)$$

In this definition, we explicitly specify the parameters in the enabledness condition to indicate their scope, i.e., the event e is enabled for all parameters p satisfying $\text{Enabled}_p(e)$. Notice that p appears freely in $\text{Enabled}_p(e)$.

4.2 Specifying Availability Properties with Rigid Events and Parameters

In Event-B, the availability of an event is determined by its enabledness condition, e.g., stating that *the event must be enabled under certain conditions*. As event guards can be strengthened during refinement, an event available in the abstraction might no longer be available in the refinement. As a result, we need a notation to specify which events' availability we are interested in, and treat them differently in the refinement. To signify that we are interested in the availability of event e, we say that event e is a *'rigid'* event, denoted as [e].

Of course, the availability of an event also needs to take into account its parameters. To indicate that we are interested in the availability of the event with respect to parameters p, we say that p are *'rigid'* parameters, denoted as [p]. Note that only rigid events can have rigid parameters.

In general, consider a rigid event e with rigid parameters rp and other (non-rigid) parameters op of the following form:[1]

> event [e]
> any [rp] op where G(rp, op) then ... end

The above rigid event means that the system satisfies the availability property stating that "event e must be enabled for any parameter rp satisfying Enabled_{rp} (e)".

4.3 Refinement Preserving Availability Properties

We now discuss the refinement of rigid events preserving the associated availability properties. In a normal Event-B refinement, the guard of a refined event can be stronger than the abstract guard (and hence restrict the event enabledness). For a rigid event, on the other hand, we want to 'preserve' enabledness, that is do not strengthen in the refinement the conditions for event enabledness, so that the availability of the event is maintained by refinement.

The syntactic rules are (1) rigid events can only be refined by rigid events, and (2) the abstract rigid parameters must be retained in the concrete events.

---

[1] We use [ ] to distinguish rigid events and parameters from others.

Consider the following abstract rigid event [ae] and concrete rigid event [ce].

event [ae]
any [rp] oap where
  Ga(rp, oap, v)
then
  // abstract action
  ...
end

event [ce]
any [rp] ocp where
  Gc(rp, ocp, v, w)
then
  // concrete action
  ...
end

Here, [rp] represents the rigid parameters, oap and ocp are respectively the other abstract and concrete parameters. While v and w are the abstract and concrete variables.

To ensure that the availability properties are preserved through refinement, we must prove that the concrete event *does not* strengthen the enableness of the abstract event, taking into account the rigid parameters. We propose the following enabledness preservation PO (denoted as ENBL),

$$I(v), J(v, w), Ga(rp, oap, v) \vdash \exists ocp \cdot Gc(rp, ocp, v, w),$$

where I(v) and J(v, w) are the abstract and concrete invariants.

In general, more rigid parameters can be introduced during refinement. Notice that the ENBL proof obligation depends on the abstract rigid parameters, i.e., any rigid parameters *newly introduced* by a refinement will be treated as non-rigid parameters for the purpose of the current ENBL PO and will only be relevant for the ENBL PO in further refinement steps.

In Event-B an abstract event can be refined by a group of concrete events. In this case, the ENBL PO is generalised accordingly. Given an abstract rigid event [ae] and concrete rigid events [ce_1], [ce_2], .., [ce_n], the ENBL PO is as follows.

$$\forall rp, oap \cdot Ga(rp, oap) \Rightarrow \bigvee_i (\exists ocp_i \cdot Gc_i(rp, ocp_i))$$

In the formula above ocp_i and Gc_i ($i \in 1 .. n$) are the other concrete parameters and guards of the corresponding concrete event ce_i.

5   SBB Systems Model in Event-B

In this section, we first present our refinement strategy and then show how we modelled the SBB system introduced in Sect. 2 using Event-B. We also illustrate the reasoning about availability properties using enabledness conditions based on the approach of Sect. 4.

5.1   Refinement Strategy

Our refinement plan consists of an abstract and four refinement levels. Although our focus is modelling the SBB, it is important to take into consideration how it interacts with other components of the system, in particular the BMD which generates the encrypted ballots for authorised voters only.

Abstract Model: We start by modelling an ideal voting system, where legitimate ballots are created for voters who have not voted before and only legitimate ballots are cast.

First Refinement: In this refinement we introduce the physical paper ballots and we distinguish between the different types of ballots according to Fig. 1. We model possible attackers behaviour, where an attacker can produce illegitimate ballots or duplicate legitimate ballots which can invalidate legitimate ballots. Voters with valid ballots will have the option to either cast or spoil their ballots. Ballots which are spoiled, invalid or illegitimate will be ejected by the SBB.

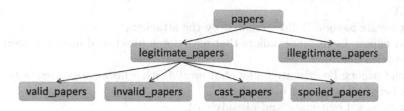

Fig. 1. Different types of paper ballots.

Second Refinement: We introduce time and invalidate ballots with expired timestamps. We assume synchronised clocks for both the SBB and BMD, but later we will show how time can be the subject of malicious attacks.

Third Refinement: We data refine the voter information and their votes by encrypting the ballots using a secret key.

Fourth Refinement: Ensuring the legitimacy of ballots is done through the MAC, where we compare the computed MAC using the secret authentication key and the MAC in the ballot barcode.

5.2   Abstract Level: Modelling an Ideal Voting System

An ideal voting system is when each voter can have at most one vote. This is ensured by the typeof−ballots invariant, where ballots are defined as a partial function between VOTER and VOTE.

@typeof−ballots: ballots $\in$ VOTER $\rightarrowtail$ VOTE

Here, we only allow the creation of ballots for voters who were not issued any ballots previously. Casting of ballots is only applicable for ballots that are not invalidated, where the variable cast represents the set of cast ballots cast $\subseteq$ ballots. Thus, we have a perfect secure voting system where we ensure that a voter can cast at most one valid ballot. At this level we only have three events create_ballot, invalidate_ballots and cast_ballot.

5.3   First Refinement: Introducing Physical Ballots and Possible Attacker Capabilities

At this level, we refine ballots and the cast ballots with the physical paper ballots (papers). These paper ballots are susceptible to attacks, e.g. duplicating paper ballots, making the voting system insecure. In this section we distinguish between the different types of papers using disjoint variables as shown in Fig. 1. Ballot papers can be either legitimate or illegitimate, and legitimate papers can be in turn partitioned as valid, invalid, cast or spoiled which are modelled as disjoint sets as follows:

- legitimate_papers: Ballots created by BMD or copied from one created by BMD.
- illegitimate_papers: Ballots created by the attackers.
- valid_papers: Legitimate ballots that have not expired, and have not been cast or spoiled before.
- invalid_papers: Legitimate ballots but invalid either because expired or (a copy has been) already cast or spoiled.
- cast_papers: Legitimate and already cast.
- spoiled_papers: Legitimate ballots that are spoiled.

We define the events that lead to creating legitimate and illegitimate ballot papers and the events that invalidate legitimate ballots. All legitimate valid ballots are created by the BMD in the event BMD_issues_paper which refines the abstract event create_ballot. We also define three possible attacks: ATK_creates_paper creates an illegitimate ballot; in this case we assume the attacker does not know the secret keys. On the other hand, both attack events ATK_duplicates_valid_paper and ATK_duplicates_invalid_paper create legitimate ballots by duplicating existing valid and invalid ballots respectively; in this case the SBB will only accept one valid copy from the voter. The abstract event invalidate_ballots refers to the ballots that cannot be cast; here we refine it by two events papers_expired and spoil_valid_ballot where any ballots that expire or get spoiled cannot be cast anymore. The abstract event cast_ballot is refined by cast_paper where only valid_papers can be cast. We also introduce eject_paper where only invalid, spoiled or illegitimate ballots are ejected out of the SBB.

Given the new variables paper_voter and paper_vote which are projection functions on papers to represent the voter's information and choices, the following invariants should hold. Invariant no_valid_double_voting_vote ensures that if two valid ballots exist for the same voter, then they are copies of each other. On the other hand, no_cast_double_voting_vote ensures that once a valid ballot is cast, the voter cannot have any other valid ballot papers. Invariants gluing_ballots and gluing_cast are gluing invariants that relate the paper ballot with the logical ballots and cast variables.[2]

---

[2] where [] is a relational image and { | } is a set comprehension. A concise summary of Event-B syntax is available at http://wiki.event-b.org/images/EventB-Summary.pdf.

@no_valid_double_voting_vote:
$\forall$b1, b2 · b1 $\in$ valid_papers $\wedge$ b2 $\in$ valid_papers $\wedge$ paper_voter(b1) = paper_voter(
    b2)
$$\Rightarrow \text{paper\_vote(b1)} = \text{paper\_vote(b2)}$$

@no_cast_double_voting_vote:
$\forall$b · b $\in$ cast_papers $\Rightarrow$ paper_voter(b) $\notin$ paper_voter[valid_papers]

@gluing$-$ballots:
ballots = {paper · paper $\in$ valid_papers $\cup$ cast_papers
                | paper_voter(paper) $\mapsto$ paper_vote(paper)}

@gluing$-$cast:
cast = {paper·paper $\in$ cast_papers | paper_voter(paper) $\mapsto$ paper_vote(paper)}

At this level, we focus at the main security properties of the SBB:

1. Accept all valid ballots.
2. Reject invalid ballots.

These two goals have different purposes, the first one is concerned with the system availability, a key security property, where we need to make sure that valid ballots are not blocked from being cast. On the other hand, the combination of both goals will ensure the integrity of voting system. The second goal is expressed in the invariant

$$\text{ejected\_papers} \subseteq (\text{spoiled\_papers} \cup \text{invalid\_papers} \cup \text{illegitimate\_papers})$$

to ensure that only rejected ballots (invalid or illegitimate) and the ballots the user choose to spoil will be ejected out of the SBB. Refinement consistency POs will ensure that this invariant holds across the different refinement levels.

The availability property can be captured by the guard of the relevant events. In particular, we specify that cast_paper is a rigid event and its parameter paper is also rigid. Using the notation introduced in Sect. 4, the event cast_paper has the following form.

event [cast_paper]
any [paper] where
@valid$-$paper: paper $\in$ valid_papers
then
   // actions for casting the ballot
   ...
end

The above event specifies that the cast_paper event must be enabled for any valid paper, i.e., satisfying the guard paper $\in$ valid_papers.

5.4   Second Refinement: Introducing Time and Availability of Events

In this refinement, we introduce Time where each ballot has a timestamp defined as paper_time $\in$ papers $\rightarrow$ TIME and current_time demonstrates the progression of time. A ballot can expire after a certain time if it is not cast or spoiled, thus invalidating a legitimate ballot. Ballots issued in the future are also considered invalid. We introduce the clock_tick event to progress time which also refines papers_expired from Sect. 5.3.

By introducing time, now we have all the conditions to define a valid ballot both in terms of duplication and expiry of time stamp. We therefore refine for the guards of the events that depend on ballot validity, such as casting and ejecting ballots. These guards will ensure that ballots have not been cast or spoiled before, have not expired nor have they been issued by an illegitimate source.

The cast_paper event, introduced at the previous level, is a rigid event with the rigid parameter paper. The refinement of the cast_paper event is as follow.

```
event [cast_paper] refines cast_paper
any [paper] where
  @typeof−paper: paper ∈ papers
  @not−already−expired: paper_time(paper) ≥ current_time − expiry_duration
  @copy−not−already−cast: paper_voter(paper) ∉ paper_voter[cast_papers]
  @copy−not−already−spoiled:
  (∀sp · sp ∈ spoiled_papers ⇒
            paper_voter(paper) ≠ paper_voter(sp) ∨
            paper_vote(paper) ≠ paper_vote(sp) ∨
            paper_time(paper) ≠ paper_time(sp)
  )
  @not−illegitimate−paper: paper ∉ illegitimate_papers
then
    // cast the paper
    ...
end
```

This gives rise to the ENBL PO at this refinement. Because ENBL PO is not yet supported by Rodin, we encode this ENBL PO as the following theorem (accept−valid−paper) in our model.

```
theorem @accept−valid−paper:  // Encoding of the ENBL PO for cast_paper
    event
∀paper ·                        // Universally quantified over abstract rigid
    variables
  paper ∈ valid_papers          // Guard of the abstract event
⇒
  /* Guards of the concrete event */
  paper ∈ papers ∧
  paper_time(paper) ≥ current_time − expiry_duration ∧
```

paper_voter(paper) ∉ paper_voter[cast_papers] ∧
(∀sp · sp ∈ spoiled_papers ⇒
 paper_voter(paper) ≠ paper_voter(sp) ∨
 paper_vote(paper) ≠ paper_vote(sp) ∨
 paper_time(paper) ≠ paper_time(sp)
) ∧
paper ∉ illegitimate_papers

Notice that there are no non-rigid parameters for cast_paper event, i.e., the right-hand side of the implication is not existentially quantified. The same applies to spoil_valid_paper because the voter should have the option to either cast or spoil their valid ballot paper, so both events have the same guards or enabling conditions and the theorem will apply to both events.

As mentioned earlier, if a ballot has been cast or spoiled before, it will be considered invalid. However, there is a difference here between the two cases, a voter with a spoiled ballot can be issued another paper if they present a physical proof to the BMD of the spoiled ballot, whereas a voter with a cast ballot cannot. Hence the difference between the guards for checking whether a paper has been cast or spoiled. In the case of a spoiled ballot another ballot belonging to the same voter can be considered valid if it at least has a different time stamp. Using model checking it was possible to discover such difference and add this assumption in the form of guard to BMD_issues_paper.

Further to the typing invariants related to the new timing variables, the following invariants describe the difference between valid and invalid ballots in regards to time and double voting and can ensure that theorem (accept−valid−paper) above is true. The last invariant valid−and−spoiled−papers−disjoint was actually discovered by attempting to prove that there is no valid ballot which is an exact copy of a spoiled ballot. A voter with a spoiled ballot can get a new legitimate ballot, but the new ballot will have at least a different time stamp (paper_time).

// The valid ballot papers must not have the future time stamps.
@no−future−valid−papers:
 ∀b·b ∈ valid_papers ⇒ paper_time(b) ≤ current_time

// The valid ballot papers must not expired
@no−expiry−valid−papers:
 ∀b·b ∈ valid_papers \ cast_papers
    ⇒ current_time − expiry_duration ≤ paper_time(b)

// For two different valid ballot papers, if it is for the same voter then
// they must have the same time stamp, i.e., they are copies of each other.
@no_valid_double_voting_time:
 ∀b1, b2 ·
  b1 ∈ valid_papers ∧ b2 ∈ valid_papers
    ∧ b1 ≠ b2 ∧ paper_voter(b1) = paper_voter(b2)
     ⇒ paper_time(b1) = paper_time(b2)

// Any invalid paper will either be expired, or a copy has been cast or a
// copy has been spoiled.
@expired−or−cast−or−spoiled−copy−invalid_papers:
  ∀paper · paper ∈ invalid_papers ⇒
    current_time − expiry_duration > paper_time(paper)
  ∨ paper_voter(paper) ∈ paper_voter[cast_papers]
  ∨ (∃sp · sp ∈ spoiled_papers
          ∧ paper_voter(paper) = paper_voter(sp)
          ∧ paper_vote(paper) = paper_vote(sp)
          ∧ paper_time(paper) = paper_time(sp)
)

@valid−and−spoiled−papers−disjoint:
  ∀vp, sp · vp ∈ valid_papers ∧ sp ∈ spoiled_papers ⇒
          paper_voter(vp) ≠ paper_voter(sp)
        ∨ paper_vote(vp) ≠ paper_vote(sp)
        ∨ paper_time(vp) ≠ paper_time(sp)

5.5   Third Refinement: Ballot Encryption

At this level we introduce encryption so tat the SBB will not be able to access
the voters information. Consequently, we apply data refinement to replace the
variables paper_vote and paper_voter with the encrypted ballot. The following
invariants describe ballot encryption and include gluing invariants to relate the
new variable paper_encrypted_ballot with the disappearing variables paper_voter
and paper_vote.

@typeof−paper_encrypted_ballot:
  paper_encrypted_ballot ∈ papers → CYPHER_TEXT

@gluing−legitimate−papers:
  ∀paper·paper ∈ legitimate_papers
          ⇒ EncryptionAlgorithm(
                  paper_voter(paper) ↦ paper_vote(paper) ↦ EncryptionKey
) = paper_encrypted_ballot(paper)

@encrypted−ballot−disjoint−cast−valid:
  ∀p · p ∈ legitimate_papers
          ∧ paper_encrypted_ballot(p) ∉ paper_encrypted_ballot[cast_papers]
          ∧ (∀sp · sp ∈ spoiled_papers ⇒
                  paper_encrypted_ballot(p) ≠ paper_encrypted_ballot(sp)
              ∨ paper_time(p) ≠ paper_time(sp))
          ∧ paper_time(p) ≥ current_time − expiry_duration
                  ⇒ paper_voter(p) ∉ paper_voter[cast_papers]

@gluing_encryption_voter:
∀ p1, p2· p1 ∈ legitimate_papers ∧ p2 ∈ legitimate_papers
    ∧ paper_encrypted_ballot(p1) = paper_encrypted_ballot(p2)
       ⇒ paper_voter(p1) = paper_voter(p2)

@inv_not_already_cast:
∀ p· p ∈ valid_papers
    ⇒ paper_encrypted_ballot(p) ∉ paper_encrypted_ballot [cast_papers]

As the SBB cannot access the voter's information on the ballot directly, the guards for cast_paper need to be refined. As a result, the ENBL PO needs to be proved for the refinement. In particular, the part of the ENBL PO related to the refinement is shown below.

theorem @accept−valid−paper:
∀paper · ... ∧
  (∀sp · sp ∈ spoiled_papers ⇒
     paper_voter(paper) ≠ paper_voter(sp) ∨ paper_vote(paper) ≠ paper_vote(
  sp)
      ∨ paper_time(paper) ≠ paper_time(sp)) ∧ ...
  ⇒ ... ∧
  (∀sp · sp ∈ spoiled_papers ⇒
     paper_encrypted_ballot(paper) ≠ paper_encrypted_ballot(sp)
     ∨ paper_time(paper) ≠ paper_time(sp)) ∧ ...

Notice that this PO is discharged trivially due to the property of the encryption function.

5.6 Fourth Refinement: Ballot Authentication

The purpose of ballot authentication is to protect against malicious intruder behaviour, where an intruder tries to cast a ballot not issued by its only legitimate source, BMD. These attacks are introduced earlier in Sect. 5.3 and specified as the attacker events. We introduce MAC to check the legitimacy of the ballot. Therefore, all the event guards checking for ballot legitimacy will be replaced by an equality check of the MAC paper_mac with the calculated MAC. The MAC can be calculated using the MACAlgorithm which requires the secret MACKey.

@typeof−paper_mac: paper_mac ∈ papers → MAC

@mac−illegitimate_papers: ∀paper · paper ∈ illegitimate_papers ⇒
paper_mac(paper) ≠ MACAlgorithm(
    paper_time(paper) ↦ paper_encrypted_ballot(paper) ↦ MACKey)

@mac−legitimate_papers:
∀paper · paper ∈ legitimate_papers ⇒
    paper_mac(paper) = MACAlgorithm(
        paper_time(paper) ↦ paper_encrypted_ballot(paper) ↦ MACKey)

The invariants mac−illegitimate_papers and mac−legitimate_papers define the difference between legitimate and illegitimate ballots in relation to MAC. As a consequence the cast_papers guards are refined to check for MAC equality and the ENBL PO is generated (similarly to the previous section).

In ATK_creates_paper, we assume the attacker does not know MACKey and will create illegitimate paper ballots. However, if an attacker compromises key, they will be able to generate ballots that are accepted by the SBB. It is therefore crucial to ensure the secrecy of this key. This could be achieved, for example, using secure hardware which can provide memory protection to the secret keys.

6   Debugging Models Using Model Checking

In this section, we discuss the use of model checking to help with debugging our model. We first discuss the analysis of refinement consistency of the rigid events in Sect. 6.1. Subsequently, we analyse the attack on the clocks of the BMD and SBB in Sect. 6.2.

6.1   Consistency of the Refinement of the Rigid Events

As presented in the previous section, the consistency of the refinement of the rigid event cast_paper is captured as the theorem accept−valid−paper. The theorem states that a valid paper will be the one that is not yet expired, the voter on the ballot has not yet voted, and a copy of the paper has not yet been spoiled. In our initial model the theorem could not be discharged automatically. ProB Model checker shows a trace that can violate the theorem as follows.

INITIALISATION
BMD_issues_paper(PAPER1, VOTER1, VOTE7)
spoil_valid_paper(PAPER1)
BMD_issues_paper(PAPER2, VOTER1, VOTE7)

The trace shows a scenario where a VOTER1 got a ballot PAPER1 with VOTE7 from the BMD, subsequently spoils the paper, and get another ballot PAPER2 with the same choice VOTE7. At this point, since PAPER2 is a valid paper, but it has the same information as the spoiled paper PAPER1 (including the time stamp), PAPER2 cannot be cast. An important assumption that is missing from our model is that the papers PAPER1 and PAPER2 must have a different time stamps. As a result, we strengthen the guard of BMD_issue_paper to add the assumption

@no−clash−spoiled−papers:
$\forall$sp · sp ∈ spoiled_papers ⇒
     voter ≠ paper_voter(sp) ∨ vote ≠ paper_vote(sp) ∨ current_time ≠
     paper_time(sp)

Furthermore, we add an invariant to state the relationship between valid papers and spoiled papers.

@valid−and−spoiled−papers−disjoint:

$\forall vp, sp \cdot vp \in$ valid_papers $\land sp \in$ spoiled_papers $\Rightarrow$
   paper_voter(vp) $\neq$ paper_voter(sp) $\lor$ paper_vote(vp) $\neq$ paper_vote(sp)
   $\lor$ paper_time(vp) $\neq$ paper_time(sp)

Given the above invariant, the theorem for proving the consistency of the refinement of the rigid event is proved automatically.

6.2 Attacks on the Clocks

In Sect. 5.4, we use one global clock and consider that both the SBB and BMD are always synchronised with current_time. However, this might not be the case and there can be some attacks on the clocks that can lead to accepting invalid paper ballots or rejecting valid paper ballots.

To model such attacks, in addition to current_time in Sect. 5.4, we introduce two clock variables BMD_time and SBB_time that are synchronised with the global clock current_time. All the invariants related to time will remain the same based on current_time, while the BMD events such as BMD_issues_paper will use the BMD_time. Similarly, the SBB events, cast_paper and eject_paper, will use the SBB_time. Therefore, to prove the invariants related to paper_time, both clocks should be equal to current_time. The equality invariants cannot be proved, if we introduce attacker events that can advance or delay the SBB and BMD clocks.

We use the ProB model checker to show how the clock attacks can invalidate the two main requirements of the case study. To help ProB automatically generate a counter example: We restrict ProB to one refinement level. We remove the invariants related to equality of time and relating the validity of ballots and time. Then, we copy over the ejected_papers type invariant that ensure that only invalid, illegitimate or spoiled ballots are ejected from previous refinement level. The model checker generates the following counter example: ⟨attacker_advance_bmd_clock(3), BMD_issues_paper(PAPER1, VOTER3, VOTE1), eject_paper(PAPER1)⟩

This trace will violate the ejected_papers type invariant copied from the first refinement, because PAPER1 $\in$ valid_papers. Therefore, using ProB, we have shown how an attack on the BMD clock can lead to rejecting a valid ballot that has not been spoiled. Other clock attacks can be demonstrated in a similar way.

7 Related Work

In this paper we have modelled a smart ballot box of a secure voting system. We use Event-B to analyse the system level security focusing on the refinement of the availability property, whereas most security verification tools such as [10,11] consider the verification of security protocols.

In [6], the authors also use a correct-by-construction approach using Event-B to model a secure e-voting system. The authors focus on the recording and the tallying phases to ensure the verifiability of the system using a decomposition pattern and a contextualisation technique. Our case study focuses on the smart ballot box which only allows the casting of valid encrypted ballots. The encrypted ballots in the SBB can in turn be used for rapid digital tabulation (tallying) and to provide an evidence-based auditing for the tabulation process. In this model we do not handle tabulation, but we are considering the extension of our models to include tabulation as a future work. Similarly, in [4], the authors focus on the verifiability of a peered web bulletin board for publishing the evidence of voting and tallying using Event-B.

In order to prove the availability of casting ballots through refinement, we prove the enabledness preservation of the events. In [12], the authors use enabledness preservation in conjunction with non-divergence to prove the liveness of an Event-B model. In their case enabledness preservation can have two notions, in the weakest notion, the enabledness preservation states if one of the events in the abstraction is enabled then one or more events in the refinement are also enabled. While, the strongest notion states if an abstract event is enabled then either the refining event is enabled or one of the new events are enabled. Even their strongest notion of enabledness preservation is still weaker than our definition of enabledness preservation which requires proving the non-strengthening of the guards of the rigid event. In [7], the authors use enabledness proofs to ensure the "refinement equivalence" of external events in the shared-variable decomposition of Event-B models. However, their proposed POs are similar to the standard Event-B POs. The idea of enabledness preservation has also been considered in the formal method ASM [3], in the concepts of ground model and refinements where the abstract and concrete guards are equivalent.

8   Conclusions and Future Work

In this paper, we have shown how the availability property of an event can be ensured through refinement by preserving the enabledness of its corresponding refined events. Such property relates to the parameter type where some parameters are considered rigid and should be preserved through refinement. We provide a general PO (ENBL) that can be applied to any event with rigid parameters. We apply ENBL PO by defining a theorem to ensure the availability of casting valid ballots.

In the future, we will focus on the semantics model to justify the soundness of the rigid property of events, we can possibly explore failure semantics. We will also look at how introducing new events in refinement can affect the ENBL PO. Finally, we plan to provide tool support for the enabledness preservation PO in Rodin. This can be done by extending the CamilleX [8] textual framework. In the CamilleX textual editor, the modeller will identify the rigid parameters of the event, and the enabledness preservation theorems will be automatically added to the Event-B generated machine.

Acknowledgement. This work is supported by the HD-Sec project, which was funded by the Digital Security by Design (DSbD) Programme delivered by UKRI to support the DSbD ecosystem.

We would like to thank Joseph Kiniry and Daniel Zimmerman from Galois for providing details of and insights into the case study.

References

1. Abrial, J.-R.: Modeling in Event-B: System and Software Engineering. Cambridge University Press, Cambridge (2010)
2. Abrial, J.-R., Butler, M., Hallerstede, S., Hoang, T.S., Mehta, F., Voisin, L.: Rodin: an open toolset for modelling and reasoning in Event-B. STTT **12**(6), 447–466 (2010). https://doi.org/10.1007/s10009-010-0145-y
3. Börger, E.: The ASM ground model method as a foundation of requirements engineering. In: Dershowitz, N. (ed.) Verification: Theory and Practice. LNCS, vol. 2772, pp. 145–160. Springer, Heidelberg (2003). https://doi.org/10.1007/978-3-540-39910-0_6
4. Culnane, C., Schneider, S.: A peered bulletin board for robust use in verifiable voting systems. In: 2014 IEEE 27th Computer Security Foundations Symposium (CSF), Los Alamitos, CA, USA, pp. 169–183. IEEE Computer Society (2014)
5. Galois and Free & Fair: The BESSPIN Voting System, May 2019. https://github.com/GaloisInc/BESSPIN-Voting-System-Demonstrator-2019. Accessed 02 Feb 2021
6. Gibson, J.P., Kherroubi, S., Méry, D.: Applying a dependency mechanism for voting protocol models using Event-B. In: Bouajjani, A., Silva, A. (eds.) FORTE 2017. LNCS, vol. 10321, pp. 124–138. Springer, Cham (2017). https://doi.org/10.1007/978-3-319-60225-7_9
7. Hallerstede, S., Hoang, T.S.: Refinement of decomposed models by interface instantiation. Sci. Comput. Program. **94**, 144–163 (2014)
8. Hoang, T. S., Dghaym, D.: Event-B and Rodin Wiki: CamilleX (2018). http://wiki.event-b.org/index.php/CamilleX. Accessed Feb 2021
9. Leuschel, M., Butler, M.: ProB: a model checker for B. In: Araki, K., Gnesi, S., Mandrioli, D. (eds.) FME 2003. LNCS, vol. 2805, pp. 855–874. Springer, Heidelberg (2003). https://doi.org/10.1007/978-3-540-45236-2_46
10. Lowe, G.: Casper: a compiler for the analysis of security protocols. In: Proceedings 10th Computer Security Foundations Workshop, pp. 18–30 (1997)
11. Schmidt, B., Meier, S., Cremers, C., Basin, D.: Automated analysis of Diffie-Hellman protocols and advanced security properties. In: 2012 IEEE 25th Computer Security Foundations Symposium, pp. 78–94 (2012)
12. Yadav, D., Butler, M.: Verification of liveness properties in distributed systems. In: Ranka, S., et al. (eds.) IC3 2009. CCIS, vol. 40, pp. 625–636. Springer, Heidelberg (2009). https://doi.org/10.1007/978-3-642-03547-0_59

Proving the Safety of a Sliding Window Protocol with Event-B

Sophie Coudert[✉]

LTCI, Télécom Paris, Paris, France
sophie.coudert@telecom-paris.fr

Abstract. This paper presents an Event-B modeling of the general version of the Sliding Window Protocol (SWP). SWPs ensure reliable data transfer over unreliable media by routing frames together with their indexes. Providing SWPs with formal guarantees is recognized to be quite complex. The experiment we present here shows that Event-B refinement is a suitable approach to ensure the safety of the protocol. First a simple model is developed with unbounded frame indexes. Then bounded indexes and modular arithmetic are introduced, as concrete indexes have fixed size. At this "hybrid" level, unbounded indexes are not used any more in computations but they are still useful to express some properties. Finally, abstract general media are refined towards queues, as an example of implementation. All unbounded indexes fully disappear in the final model.

Keywords: Event-B · Sliding Window Protocol · Formal refinement · Safety

1 Introduction

This paper experiments the use of the Event-B approach [1] to ensure the safety of a Sliding Window Protocol (SWP). SWPs are a well-known family of communication protocols that ensure a reliable communication from a sender to a recipient over unreliable media [2,3]. They are widely used and in particular in TCP (Transmission Control Protocol) and HDLC (High-Level Data Link Control). Unreliable media may loss, re-order or duplicate messages. SWPs overcome this problem by implementing an acknowledgment medium in addition to the frame medium, and by transporting frames together with their index, which allows re-ordering at receiver side. Moreover, to offer high latency, they implement windows at both sides of communication. It allows to avoid waiting for the acknowledgment of a frame before sending the next one. There are several versions of the protocol, depending on the window sizes. We modeled the most general version, as the sending and receiving window sizes are parameters of our specification.

Despite the relative simplicity of the protocol, it is widely recognized that ensuring its correctness is far from obvious, because of strong parallelism and

© Springer Nature Switzerland AG 2021
A. Raschke and D. Méry (Eds.): ABZ 2021, LNCS 12709, pp. 50–65, 2021.
https://doi.org/10.1007/978-3-030-77543-8_4

subtle interactions between components. Many previous research propose formal and non-formal modeling and proofs for SWPs. They rely on various techniques such as model checking [4–6] or deductive proof [7–9]. The proposed solutions are generally complex or limited in scope, and no ideal approach emerges. Here we experiment the use of Event-B. Using Event-B, systems are described by way of guarded events. Properties about described state machines are ensured by interactive proofs of generated proof obligations. The main feature of Event-B is stepwise refinement, which mathematically links abstract models to more detailed ones. It makes it possible to prove properties at abstract level (thus simple). Proved refinement then ensures that these properties hold at more concrete levels. Event-B is generally dedicated to system design and some previous works propose general approaches using it for developing distributed systems [10,11]. Here we explore the use of Event-B more as a proving technique for a well-known quite difficult problem, as the refinement process is developed *a posteriori* and guided by proving objectives.

As often in Event-B, we ensure safety but do not consider fairness assumptions nor liveness properties. Our approach can be summarized as follows. We model the safety property which says that frames are delivered in the order they have been sent. We refine it to a model of the protocol where frame indexes are unbounded values. We then introduce bounded indexes, as concrete implementations use fixed-size for them. This leads to a hybrid model where both kind of indexes co-exist. Bounded ones are used in all algorithmic aspects. Unbounded ones are still useful to express some conditions that avoid ambiguities (as a bounded index may correspond to several unbounded ones). At this level, media may still re-order frames. Finally, as an example, we refine media to lossy queues. Respecting some constraints on parameters, queues ensure non-ambiguity and allow to fully eliminate unbounded indexes. The provided models could be then refined towards more concrete implementations.

The paper is organized as follows. Section 2 briefly presents Event-B. Section 3 presents the modeled protocol, the safety property and its refinement to the model with unbounded indexes. Section 4 presents the introduction of modular arithmetic and the refinement of media towards queues with bounded indexes. Section 5 presents related works before concluding.

2   Event-B in Brief

Using the formal method Event-B [1,12], systems are described as machines including *events* modifying a *state* characterized by *variables*. Each event has a *guard* expressing the conditions in which it may occur, and a *substitution* describing how it modifies the state. Initialization is a particular event without guard. Events may be non-deterministic: for example, after substitution "$x :\in \{1, 2\}$", x's value may be 1 or 2. Then, machines characterize systems in a discrete way, by all their possible behaviours, i.e. allowed sequences of events beginning with initialization. Invariants are properties that must hold in any state. Tools generate proof obligations that ensure invariants. Intuitively, one must prove that initialization establishes the invariants and that any other event preserves them. The logic used for formulas is first-order logic enriched with elementary set theory.

Beyond invariants the main strength of the method is proved refinement which mathematically links abstract and detailed models and guarantees that behaviour and results of abstract level are respected at concrete level. More precisely, a refinement is a new machine which provides a more detailed description of the system. It includes a "gluing" invariant which links the abstract and concrete states. It refines abstract events w.r.t. the new state representation. It may also add new events which implicitly refine a "skip" at abstract level. Then each concrete event has a corresponding one in the abstract machine, which links concrete behaviours with abstract ones. Tools generate proof obligations that guarantee the correctness of the refinement. Roughly speaking, a refinement is correct if any low level behaviour corresponds to a high-level one. It both ensures that low level behaviour respect abstract one and that high level invariants are preserved, as no new behaviour appears. Intuitively, the main constraints that a refinement must respect are the following ones.

– New events cannot modify abstract state. They only detail steps on new variables that appear in the refinement.
– Refined events have stronger guards than their associated abstract event. So they can't occur when abstract event can't.
– Refined events can only increase determinism: they modify the state in a way that was allowed by the abstract event. "$x :\in \{1,2\}$" may become "$x := 2$"

In short, removing new events and variables from concrete behaviours must lead to possible abstract behaviours. For example, new events may detail steps of a calculus presented as atomic at abstract level. New variables may receive intermediate results that do not exist at abstract level. Event-B has two associated tools Rodin and atelier-B. The work we present used Rodin [12].

3   Modeling the Protocol with Unbounded Indexes

The model[1] includes 13 machines implementing 12 refinement steps. Thus, only an excerpt is presented here. The first machine describes the "user view" of a safe communication. The second one introduces the windows. The third one introduces the communication media. Then a technical step prepare the introduction of bounded indexes. All these steps are simple and provide a safe model of the SWP considering unbounded indexes and an abstract description of media.

3.1   The Sliding Window Protocol

We modeled the most usual general version of the sliding window protocol, with a frame medium from transmitter to receiver, and an acknowledgment medium from receiver to transmitter. Both media are unreliable.

---

[1] https://perso.telecom-paristech.fr/coudert/downloads/SWProdin3_5v1.zip.

The transmitter has a window of size wt with lower bound tw. It can transmit all frames with index in interval $tw \ldots tw+wt-1$ before receiving an acknowledgment for frame index tw. Acknowledgment are cumulative: receiving acknowledgment i at the transmitter side ensures that all frames with index $i' < i$ have been delivered at the receiver side. Thus, each time the transmitter receive an acknowledgment i that is greater than tw, it updates this lower bound ($tw := i$) and the window slides. To avoid waiting indefinitely for an acknowledgment, the transmitter implements timeouts. Relying on these timeouts, frames in the window may be retransmitted. Thus they are memorized. The precise mechanism of timeouts depends on implementation. Our model relies on the general possibility of retransmission, which is sufficient to ensure safety.

The receiver may accept frames in the wrong order. Thus it also implements a window, with lower bound rw and size wr. rw is the index of the lowest frame not yet accepted. Each time a frame arrives the receiver computes the new value for rw and sends an acknowledgment with this value. If the frame is outside the window, it is discarded. Otherwise it is put in the window. rw is updated when the frame index is $rw + 1$. When this happens, all the frames between the old and the new value of rw have been accepted and correctly ordered. They are removed from the window and put in a FIFO buffer which contains the frames that can be delivered to the recipient, in the correct order.

For historical reasons, this version differs slightly from the standard one: the receiver FIFO is not required by the protocol definition. This difference is not a simplification. Considering the abstraction of our intermediate hybrid model, media can duplicate, lose and re-order messages. The refinement towards queues only considers losses. Implementation of media must avoid ambiguity. Indeed, bounded indexes have several associated unbounded ones. When using queues non-ambiguity is ensured provided that $wt + wr \le m$, where m is the modulus. This is proved in our modeling. When media re-order frames, mechanisms based on frame lifetime are used. These mechanisms are various and often complex (in particular for TCP). Moreover, handling time with Event-B is not so simple [13,14]. Thus we delay this aspect to future works.

3.2   Safety Property: Behaviour of Reliable Communication

The expected behaviour is the following one: any transmitted frame is received once and the order is respected. In other words, *the sequence R of received frames is a prefix of the sequence T of transmitted frames* (in Fig. 1).

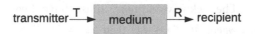

Fig. 1. Communication scheme

Our most abstract Event-B machine simply describes this. Frames belong to an abstract set Fr. Variable t is the unbounded index of the next frame to send.

Variable r is the unbounded index of the next frame to receive. Sequences are modeled by total functions in a usual way and the property is a simple inclusion. The corresponding invariants are:

$$T \in 0 .. t - 1 \rightarrow Fr \quad \text{and} \quad R \in 0 .. r - 1 \rightarrow Fr \quad \text{and} \quad R \subseteq T$$

Initial sequences are empty and initial indexes are null. To obtain a model representing the expected transmission behaviour, we define two events which trivially preserve invariants:

> $-$ Event transmit $\widehat{=}$ any f where $f \in Fr$
>   then $T := T \cup \{t \mapsto f\} \parallel t := t + 1$
> $-$ Event receive $\widehat{=}$ when $t > r$ then $R(r) := T(r) \parallel r := r + 1$

A received data must have been sent ($t > r$), and it is the good one ($T(r)$).

3.3 Introducing Windows and Receive Buffer

The first refinement step introduces the windows. Constants wt and wr are respectively the (strictly positive) sizes of the transmitting and receiving windows. Variables tw and rw are their lower bounds. Figure 2 illustrates the state and events of this refinement. In formulas, variable i is reserved for unbounded indexes and variable f is reserved for frames.

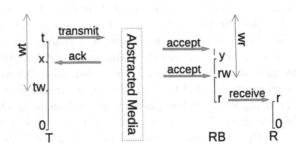

Fig. 2. First refinement

A window constraint $t < tw + wt$ is added to the guard of the transmit event.

> $-$ Event transmit $\widehat{=}$ any f where $f \in Fr \wedge t < tw + wt$ then $\ldots$

New event "ack" models the delivering of an acknowledgment inside the transmitter window, which makes this window slide: tw becomes x on the left side of Fig. 2. The acknowledgment must have been sent ($i \leq rw$).

> $-$ Event ack $\widehat{=}$ any i where $tw < i \leq rw$ then $tw := i$

A receiving buffer RB is introduced (right side of Fig. 2). It contains the frames in the receiving window (in $rw .. rw + wr - 1$), and those that have no yet been delivered to the recipient (in $r .. rw - 1$). $RB \subseteq T$ and $r .. rw - 1 \subseteq dom(RB)$ are invariants. The event receive now transfers frames at index r from RB to[2] R.

> – Event receive $\widehat{=}$ when $r \in dom(RB)$
> then $R := R \cup (\{r\} \lhd RB) \parallel RB := \{r\} \lhd\!\!\!- RB \parallel r := r + 1$

The new event "accept" fills RB by accepting frames in the receiving window. The lower bound rw of this window is the first "not yet accepted" frame (thus $rw \notin dom(RB)$). This bound may change: in Fig. 2, rw becomes y for the lower "accept" arrow.

> – Event accept $\widehat{=}$ any i where $i \in (rw .. (rw + wr - 1)) ...$
> then $RB := RB \cup (\{i\} \lhd T) \parallel rw := new\_rw$
> new_rw is $max(\{i' | i' \in rw .. rw + wr \wedge rw .. i' - 1 \subseteq dom(RB) \cup \{i\}\})$

Many proofs in this refinement are either automatic or easy. The more complex aspect is the set theoretic definition of new_rw above, which also makes some proofs in next refinements tedious. At this level, communication media are fully abstracted. Only guards of events characterize the messages that may be delivered. For example, only acknowledgment of accepted message may occur, although their sending is not modeled. The next refinement provides an explicit modeling of the media.

3.4   Introducing Communication Media

The second refinement introduces the two media: TM, which routes data from transmitter to receiver, and RM, which routes acknowledgments from receiver to transmitter. It is summarized in Fig. 3. Both media are approximation of multisets by sets. Intuitively, $msg \in TM$ means that there is at least one occurrence of message msg in the frame medium. In the real world, there may be multiple occurrences when the transmitter retransmits a message and also when the network duplicates messages. Sets may represent unordered transmissions but also ordered ones, which are just a special case.

> – TM contains transmitted frames with their unbounded indexes.
> Thus $TM \subseteq T$ is an invariant.
> – RM contains acknowledgments that have been sent.
> Thus $RM \subseteq 0..rw$ is an invariant.

The event receive is unchanged. Other refined events put messages in media and get messages from media. Delivered messages may be removed from media or not (because multiple occurrences are possible).

> – transmit event: substitution $TM := TM \cup \{t \mapsto f\}$ is added.
> – accept event: substitutions $RM := RM \cup \{new\_rw\}$ and
> $TM :\in \{TM, TM \setminus \{i \mapsto f\}\}$ are added.
> any i d where $... i \mapsto f \in TM$ then $... RB := RB \cup \{i \mapsto f\}$
> replaces any i where $... i < t$ then $... RB := RB \cup (\{i\} \lhd T)$.
> – ack event: the guard $i \leq rw$ becomes $i \in RM$.

---

[2] $S \lhd F$ restricts the domain of function F to S. $S \lhd\!\!\!- F$ restricts it to $dom(F) \backslash S$.

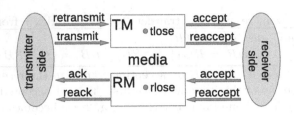

Fig. 3. Second refinement

Five new events appear, which modify media but not the abstract state of previous level. Then the inventory of events defined by the protocol is complete. Events tlose and rlose model media behaviour: they can lose or duplicate messages. With set approximation, only losses may modify state representation.

- Event tlose $\hat{=}$ any TM' where $TM' \subseteq TM$ then $TM := TM'$
- Event rlose $\hat{=}$ any RM' where $RM' \subseteq RM$ then $RM := RM'$

Event retransmit models retransmission which can occur after timeout: a frame in transmitter window is put in TM. Event reaccept models the delivery of a frame that is not in receiving window[3]: acknowledgment rw is put in RM and the frame may be removed from TM. Reminder that rw is the lower bound of the receiver window and not the index of the received frame. Event reack models the delivery of an old acknowledgment ($i_{ack} \leq tw$). This acknowledgment may be removed from RM.

All the proofs of this refinement step are either automatic or very simple. The result is an abstract but fairly complete modeling of the sliding window protocol with a proof of its safety. Many usefull invariants have been proved (not enumerated here) such as for example $tw \leq rw$, $tw \geq rw - wt$,...Notice that the sending window has not been modeled by an explicit buffer, as it is not a significant step at this level. Sent frames in this window are simply a sub-sequence of T. Further refinement could concern the media and would be different, depending on medium choices.

In our approach, we first focus on the introduction of modular arithmetic, keeping the generality of media description. The next refinement only makes a little technical change to prepare this. Bounded values will appear for indexes, in particular in the receiver window. The unbounded size of buffer RB is then a problem. The third refinement step splits RB into two parts: the receiver window RW and the receiver queue RQ. RW contains frames in $rw .. rw + wr - 1$, whose size is bounded by wr. RQ contains all frames with indexes in $r .. rw - 1$ and is not bounded. Formally, we have $RW = r .. rw - 1 \lhd RB$ and $RQ = r .. rw - 1 \lhd RB$ as gluing invariants. Event descriptions are slightly modified to accommodate this change. For example, the substitution of event

---

[3] Notice that frame outside the receiver window may be too old or too recent ones.

accept contains $RQ := RQ \cup (rw \mathbin{..} new\_rw - 1 \lhd (RW \cup \{id \mapsto f\}))$. Most manual proofs concern the accept event, and are simple.

4   Modeling the Protocol with Bounded Indexes

A first big stage leads to an hybrid model in which all computations are based on bounded indexes, but unbounded indexes are still used to express some constraints. It is the most consistent part of the work and is decomposed in three refinement steps. This intermediate model can be refined toward different kind of media. The last stage refines it towards lossy queues, making unbounded indexes disappear. It is decomposed in 6 refinement steps. Three of these steps are a bit tedious as handling queues is not trivial. The three other ones, which eliminate intermediate structures, are simple.

4.1   A Small Library for Modular Arithmetic

As the build-in modulo operator of Rodin didn't meet our needs, we defined our own modulo based on the euclidian division. The result is a fully proved collection of definitions and theorems in context machines. Most theorems are intermediate steps to finally prove well-known properties on modulo, such as congruences for example. We just show a glimpse of them here. The base of our definition is the following one, for a modulus $m \geq 1$:

- $eDR(m) = \{x \mapsto (d \mapsto r) | r \in 0 \mathbin{..} m - 1 \land x = d * m + r\}$
- $eDR(m)(x) = ediv(m)(x) \mapsto emod(m)(x)$

We proved that $eDM(m)$ is a total function. Thus $emod$ and $ediv$ are well-defined. Notice that the euclidian division is not the build-in one: $-5/2 = -2$ but $ediv(-5)(2) = -3$. We proved theorems as for example:

- $\forall m, x \cdot m \geq 1 \Rightarrow x = (ediv(m)(x)) * m + emod(m)(x)$
- $\forall m, x \cdot m \geq 1 \Rightarrow emod(m)(emod(m)(x)) = emod(m)(x)$
- $\forall m, \cdots emod(m)(x1) = emod(m)(y1) \land emod(m)(x2) = emod(m)(y2)$
  $\Rightarrow emod(m)(x1 + x2) = emod(m)(y1 + y2)$

This general library is used to prove some more high level properties required by our refinement, presented in next sections. Using universally quantified theorems in proofs is a bit tedious as it requires manual instantiation and it strongly compromises automation. Inference rules would have been more convenient[4].

---

[4] We are currently transposing them in the new release of the Rodin theory plugin.

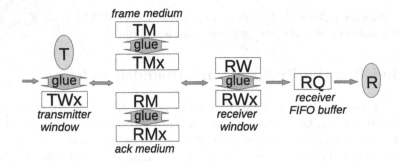

Fig. 4. Hybrid structures

4.2 Introducing Bounded Values in the Model

In concrete systems, the indexes of frame are stored with a fixed size s. Thus bounded values "modulo m" are used, where the modulus m is usually 2^s. In other words, $emod(m)(i)$ replaces i. We do not directly replace unbounded indexes but we first make both information exist simultaneously in "hybrid" structures. It allows to express properties that are useful for the proofs. A complete refinement makes unbounded indexes disappear, as they are absent from real implementations. This is the purpose of Sect. 4.3, where media are refined. Here, they occur in properties required to avoid ambiguity when delivering messages. The way these properties are ensured depends on media choices and cannot be modeled at this abstraction level. Thus unbounded indexes do not disappear but we ensure that they are no longer used for any computation purpose. Introducing modular arithmetic was rather complicated. We give an insight of the most relevant aspects of the process. In the model, the modulus m is a constant. In formulas, z is reserved for bounded indexes.

As shown on Fig. 4, a hybrid structure Sx is introduced for each abstract structure S in $\{TM, RM, RW\}$. Hybrid structures just add bounded indexes and are all linked to abstract structures by similar gluing invariants. For example, TM contains tuples $i \mapsto f$, which becomes $i \mapsto (emod(m)(i) \mapsto f)$ in TMx. An explicit transmitter window TWx appear and is linked to T. The model contains among others the following invariants or theorems.

$$TMx \subseteq 0 \mathinner{.\,.} t - 1 \times (0 \mathinner{.\,.} m - 1 \times Fr)$$
$$\forall i, z, f \cdot i \mapsto (z \mapsto f) \in TMx \Rightarrow z = emod(m)(i)$$
$$TM = \{i \mapsto f \mid \exists z \cdot i \mapsto (z \mapsto f) \in TMx\}$$
$$\dots$$
$$\forall i, z, f \cdot i \mapsto (z \mapsto f) \in TWx \Rightarrow i \mapsto f \in T$$

New variables tz, rwz and twz are added, respectively containing bounded values of t, rw and tw. Events are enriched to handle the new variables while preserving gluing invariants. During the process, all unbounded variables progressively disappear from the computation of bounded values. For example in the following excerpt, only computations of unbounded data rely on unbounded variable t.

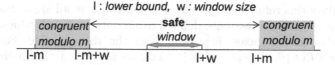

Fig. 5. Safe area w.r.t ambiguity

invariants $tz = emod(m)(t)$
— Event transmit $\hat{=} \ldots$ then $\ldots t := t + 1$
$$\| \; tz := emod(m)(tz + 1)$$
$$\| \; TMx := TMx \cup \{t \mapsto (tz \mapsto f)\} \ldots$$

This refinement required lots of manual proofs, in particular when using theorems of Sect. 4.1. But few proofs were actually difficult. The most tedious was to refine the computation of the new lower bound of the receiver window (in event accept, $new\_rw = max(complexexpressionusingrw)$), which required several intermediate steps.

The most sensible point was to avoid ambiguity when accepting a frame (or an acknowledgement). The guard "any i where $i \mapsto f \in TM \wedge i \in rw..rw+wr-1$" becomes[5] "any i z where $i \mapsto (z \mapsto f) \in TMx \wedge z \in emod(m)[rw..rw+wr-1]$". If TMx also contains $i' \mapsto (z \mapsto f')$ with $i' \neq i$, there is a risk of confusion between f and f'. Safety is lost. *Event "accept" must not occur in such a situation.* Figure 5 illustrates the constraint that must be respected by unbounded indexes, w.r.t a receiving window.

Indexes in grey intervals have the same associated bounded value as indexes in the window and must be thus excluded. Area between grey intervals is safe. This has been formally proved in a high level theorem of our context machines for modular arithmetic. Ideally, the following constraint should be an invariant of the specification:

$$\forall i, z, f \cdot i \mapsto (z \mapsto f) \in TMx \Rightarrow i \in rw - m + wr .. rw + m - 1 \qquad (1)$$

But the mechanisms that ensure it depend on the kind of medium. When frame lifetimes are used, all frame in the frame medium have different unbounded indexes. It is not necessary the case when queues are used, thanks to the preserved order. The hybrid model covers both kind of medium. Thus the mechanisms ensuring (1) are not modeled. Invariant (1) will only be provable in further refinements. Here, (1) is added to the guard of the event "accept". We can then prove that the relation $i = rw + emod(m)(z - rwz)$ holds for any tuple $i \mapsto (z \mapsto f)$ either being accepted or already in the receiving window. Thus any ambiguity at the receiver side is excluded. We follow a similar approach at the transmitter side.

---

[5] $emod(m)[rw..rw+wr-1]$ becomes $emod(m)[rwz..rwz+wr-1]$ later in refinement.

The result of this refinement process is a model where all algorithmic aspects of the sliding window protocol are defined using only bounded values. We begin to make unbounded indexes disappear from the receiver side. For this, the receiver queue RQ (in $r .. rw - 1 \rightarrow Fr$) evolves towards a "normalized" queue RQq with $RQq \in 0 .. card(RQq) - 1 \rightarrow Fr$. We also replace r with $card(R)$, which is mostly symbolic. Unbounded indexes could also be removed from the transmitter and receiver windows but it would compromise further refinement as some useful properties using them would disappear. Thus this operation is delayed.

The only remaining role for unbounded indexes is to exclude ambiguity in the guards of the accept and the ack events. They will be removed from the guards later when becoming invariants of the whole model. Indeed, in order to have a working protocol, these properties should hold anytime. Media must be refined before completely eliminating unbounded value from the specification. When media don't preserve the order of messages, message lifetimes are used. When media preserve order it suffice to have $m \geq rw + tw$. This is proved in the next section which presents the refinement of both media towards queues.

4.3   Refining Media Towards Lossy Queues

Abstract media are refined towards queues in a process with multiple steps, which can be summarized in four big stages. The first stage introduces hybrid queues $(TMxq,RMxq)$ and proves that they refine abstract media (TMx,RMx). The second stage does not modify anything but proves that non-ambiguity properties are invariants. Thus these properties are removed from the guards of events, and specification of events do not need unbounded indexes any more. The third stage introduces final structures $(TMzq,RMzq,TWz, RWz)$, i.e. representations of structures with only bounded values. The last stage removes all intermediate structures, i.e. all representations with unbounded values.

The following lines are a glimpse of the first stage, at the receiver side. Intuitively, the last line is $TMxq := tail(TMxq)$.

---

invariants $TMxq \in 0 .. card(TMxq) - 1 \rightarrow \mathbb{Z} \times (\mathbb{Z} \times Fr)$
         $TMx = ran(TMxq) \ldots$
Event accept $\widehat{=}$ any $i\ z\ f \ldots$
     where $0 \mapsto (i \mapsto (z \mapsto f)) \in TMxq$
           $z \in emod(m)[rwz .. rwz + wr - 1] \ldots$
     then $\ldots$ $TMxq := \{j \cdot j \geq 0 \wedge j + 1 \in dom(TMxq)|j \mapsto TMxq(j + 1)\}$

The following lines are a glimpse of the third stage. The witness "with $0 \mapsto (i \mapsto (z \mapsto f)) \in TMxq$" removes the unbounded index i from the list of parameters. It automatically disappears in the next refinement step.

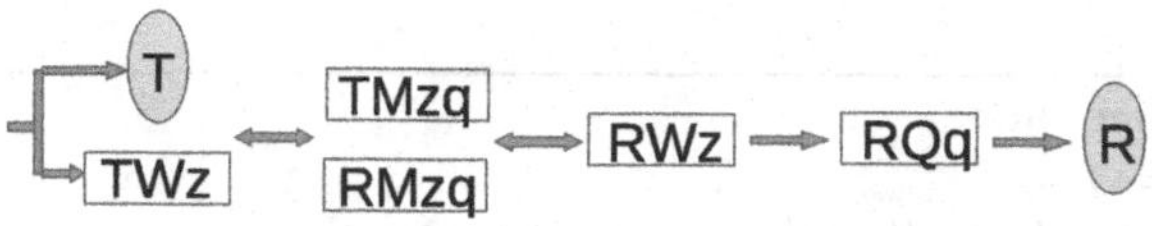

Fig. 6. Final model

> invariants $TMzq = \{j, z, f \cdot \exists i \cdot j \mapsto (i \mapsto (z \mapsto f)) \in TMxq | j \mapsto (z \mapsto f)\}$ ...
> Event accept $\widehat{=}$ any z f ...
>   where $0 \mapsto (z \mapsto f) \in TMzq$ ...
>   with $0 \mapsto (i \mapsto (z \mapsto f)) \in TMxq$ then ...

Stages one and three are a bit tedious as handling queues in proofs is not automatized, but they are essentially technical and not hard. Stage four is also technical. It eliminates abstract variables in a usual way and is simple. The most significant stage is the second one which proves the properties that make the protocol safe, provided that $m \geq wt + wr$. Here is the full list of intermediate results (invariants or theorems) which have been proved to establish the property required by the receiver side. Theorem of line 5 is the formula (1) p. 10.

1− $\forall \ldots j \mapsto (i \mapsto f) \in TMxq \Rightarrow i < t$ (theorem)

2− $\forall \ldots j_1 \mapsto (i_1 \mapsto f_1) \in TMxq \wedge j_2 \mapsto (i_2 \mapsto f_2) \in TMxq \wedge$
    $j_2 > j_1 \Rightarrow i_2 > i_1 - wt$

3− $\forall \ldots i_1 \mapsto zf \in RWx \wedge j \mapsto (i_2 \mapsto f) \in TMxq \Rightarrow i_2 > i_1 - wt$

4− $\forall \ldots j \mapsto (i \mapsto f) \in TMxq \Rightarrow i \geq rw - wt$

5− $(\forall \ldots i \mapsto (z \mapsto f) \in TMx \Rightarrow i \in rw - m + wr \mathbin{..} rw + m - 1)$ (theorem)
    More intuitively, proving line 5 this way consist in

- proving that $i \leq rw + m - 1$, i.e. $i \leq rw + wt + wr - 1$ (as $m \geq wt + wr$).
  Line 1 ensure that frames in medium have been sent, thus $i < tw + wt$;
  as $wr \geq 1$ (axiom) and $tw \leq rw$ (high level invariant), $tw \leq rw + wr - 1$
  holds, thus $tw + wt \leq rw + wt + wr - 1$ holds.
- proving that $i \geq rw - m + wr$, i.e. $i \geq rw - wt$ (as $m \geq wt + wr$). Line 2 is the
  key point, relying on order. It ensures that if i_2 is after i_1 in the queue, then
  i_2 is not older than $i_1 - wt$. Then (line 3), as soon as i_1 has been accepted,
  no frame in the queue is older than $i_1 - wt$. Thus line 4 holds (as $rw - 1$ has
  been accepted) which directly implies the required property.

So refining media towards queues is not difficult. The final model is quite simple (3 pages, with comments). Its state is summarized on Fig. 6. As an illustration, here is the complete specification of events ack[6] and tlose. Notice that now, tlose removes one element from the queue, at any place.

---

[6] Reminder: acknowledgments outside the window are handled by event reack.

Event ack $\hat{=}$ refines ack
  when $RMzq \neq \varnothing$
       $RMzq(0) \in (emod(m))[twz + 1 .. twz + wt]$
  then $twz := RMzq(0)$
       $TWz := emod(m)[twz .. twz + emod(m)(RMzq(0) - (twz + 1))] \lhd TWz$
       $RMzq := \{j \cdot j \geq 0 \wedge j + 1 \in dom(RMzq)|j \mapsto RMzq(j + 1)\}$
Event tlose $\hat{=}$ refines tlose
  any k where $k \in dom(TMzq)$ then $TMzq :=$
  $(0 .. k - 1 \lhd TMzq) \cup \{j \cdot j \geq k \wedge j + 1 \in dom(TMzq)|j \mapsto TMzq(j + 1)\}$

Further refinement should introduce concrete computations for modular arithmetic. It would involve implementation dependent choices (in particular w.r.t. hardware/software choices). Here we stay at an abstraction level where their characterization is set theoretic.

5   Related Work

Numerous previous works address the verification of SWPs, so we only mention some of them. We didn't find recent significant contribution. An interesting history is provided in [9]. Approaches based on model checking [4–6] consider fixed size windows and experience the problem of state explosion. They generally allow to handle liveness. Our work is more related to deductive approaches.

One of the first paper introducing SWP, [2], already provided an informal proof. Several more formal works address the safety of various versions of the protocol. [7] is a partially automated approach of the "Go-back-N" version (receiver window size is 1) with unbounded frame indexes. It relies on the decidable logic WS1S and requires some manual abstractions. [15] ensures the safety of the general version using PVS, a prover for typed higher order logic. An operational description of the protocol is verified using an invariant strengthening approach. Indexes are unbounded and media are FIFOs. Liveness is handled using a PVS theory for "runs". [8] also uses PVS to prove the safety of the "Selective-Repeat" version (the sizes of both windows are equal). Bounded indexes are handled using lemmas provided by PVS. The work took about 4 months, with automated proofs. It also proposes a specific handling of frame lifetime. More recently, [16] presents a similar work for the general version of the protocol, with unbounded indexes and media that do not re-order messages. [9] seems to be one of the most consistent work on the subject. Both safety and liveness (under fairness assumption) are addressed, considering the selective-repeat version of the protocol, FIFOs and indexes modulo $2n$, where n is the size of windows. The paper presents a (complex) manual proof, which is then formalized using PVS. It relies on μCRL (a process algebra with datatypes) and branching bisimulations which ensure some reachability properties. It is a general approach for protocols [17] which can be instantiated by specific protocols, such as the sliding window protocol.

We used a stepwise refinement approach. From this point of view, the works we found in the literature that are the closest to ours are [1, 18–20]. [1, 18] deal with the alternating bit protocol (window sizes are 1). [20] briefly presents a modeling of the SWP as an application of a development pattern for Event-B. The model ensures safety and considers unbounded indexes. [19] deals with a slighly modified version of the general one, considering unbounded indexes and FIFO media. Although the approach is not truly formalized, it is methodical and addresses the design of distributed algorithms. It consists of a sequence of sequential program transformations that preserve correctness. The first program is global (i.e. not distributed) and similar to our abstract machine. In a transformation step, variables are partitioned w.r.t. processes, which makes the need of communication appear, as processes can only access their own variables. New processes and variables are thus introduced, together with invariants that link them. Resulting steps for the SWP example look like our steps in Sect. 3. The approach can be compared to an Event-B refinement for the purpose of applying shared event decomposition [10, 21, 22]. Beside that, our model could be decomposed using this technique. In addition, [19] ensures a progress property using variants. It is a kind of liveness which relies on non-cumulative (thus non-standard) acknowledgments. Our approach, by comparison, is fully formalized and proved in Event-B, and considers bounded indexes. Moreover, thanks to the stepwise refinement approach, we identify different significant models: the model with unbounded indexes of Sect. 3, the hybrid model of Sect. 4.2, and the model with queues of Sect. 4.3. The two last ones could be further refined towards safe implementations.

Event-B approaches dedicated to distributed systems [10, 11] also start with a global model and introduce distribution and communications later in refinement. They aim at proposing general methodologies for the development of distributed algorithms. They do not focus on communication protocols. Lastly, [13] recently models the bounded retransmission protocol (a timed Alternating Bit Protocol) as an example. This work provides Event-B with infinite traces semantics, which allows to handle progress with respect to time. New proof obligations are identified for handling some real time and fairness properties.

6 Conclusion

In this paper we ensure the safety of the general version of the Sliding Window Protocol using the Event-B refinement approach. 326 proofs among the 794 ones where manual, but time is more significant as complexity of proofs varies widely. This work took about two and a half months to a moderately experienced Rodin user, despite some blunders in the modeling.

The refinement process is decomposed in three main significant stages. The first one quickly produced a model with unbounded frame indexes. The second one transposed all algorithmic aspects into modular arithmetic for bounded indexes. In the resulting model, unbounded indexes are only used to express properties. This general model may then be refined w.r.t. specific media. As an example, the third stage is a refinement that replaces abstract media by queues. Unbounded indexes completely disappear. First and third stages each took a long week, as did the development of a small library for modular arithmetic. Much of the effort was expended on switching between unbounded and bounded indexes. It is not surprising as avoiding ambiguity while doing this is a delicate issue. Moreover, as pointed in Sect. 4.1 having inference rules for modular arithmetic would have been more convenient.

Nearly all models are simple. Only the hybrid one, which merges both kinds of indexes, is less readable. As a final model, it could be strongly simplified. But lot of information is kept in order to make further refinement easier (and even feasible), and to avoid some useless proofs. From a methodological point of view, it is a delicate point as this model is probably the main one. Indeed, it is intended to be the starting point of any refinement towards concrete media.

To conclude, the experiment is a success and the Event-B approach has been very suitable for the problem under study. The only drawback is that it does not allow to handle liveness. However liveness is not a default property of this protocol, as messages may be lost. It requires some restrictions or additional assumptions (e.g. fairness). The absence of deadlock has been visually checked in our final model. Further work could target the refinement of our hybrid model towards media with frame lifetime mechanism. Applying the approach to other non trivial protocols would also be interesting.

References

1. Abrial, J.-R.: Modeling in Event-B: System and Software Engineering, 1st edn. Cambridge University Press, New York (2010)
2. Stenning, N.V.: A data transfer protocol. Comput. Netw. **1**(2), 99–110 (1976)
3. Tanenbaum, A.S., et al.: Computer Networks. Prentice-Hall (1996)
4. Richier, J.-L., Rodriguez, C., Sifakis, J., Voiron, J.: Verification in XESAR of the sliding window protocol. In: IFIP WG6.1 Seventh International Conference on Protocol Specification, Testing and Verification VII, NLD. North-Holland Publishing Co (1987)
5. Kaivola, R.: Using compositional preorders in the verification of sliding window protocol. In: Grumberg, O. (ed.) CAV 1997. LNCS, pp. 48–59. Springer, Heidelberg (1997). https://doi.org/10.1007/3-540-63166-6_8
6. Godefroid, P., Long, D.E.: Symbolic protocol verification with queue BDDs. Formal Methods Syst. Des. **14**(3), 257–271 (1999)
7. Smith, M.A., Klarlund, N.: Verification of a sliding window protocol using IOA and MONA. In: FORTE/PSTV 2000, pp. 19–34. NLD (2000). Kluwer, B.V
8. Chkliaev, D., Hooman, J., de Vink, E.: Verification and improvement of the sliding window protocol. In: Garavel, H., Hatcliff, J. (eds.) TACAS 2003. LNCS, vol. 2619, pp. 113–127. Springer, Heidelberg (2003). https://doi.org/10.1007/3-540-36577-X_9
9. Badban, B., Fokkink, W., Groote, J.F., Pang, J., van de Pol, J.: Verification of a sliding window protocol in μCRL and PVS. Formal Aspects Comput. **17**(3), 342–388 (2005)

10. Siala, B., Bhiri, M.T., Bodeveix, J.-P., Filali, M.: An event-B development process for the distributed BIP framework. In: Ogata, K., Lawford, M., Liu, S. (eds.) ICFEM 2016. LNCS, vol. 10009, pp. 313–328. Springer, Cham (2016). https://doi.org/10.1007/978-3-319-47846-3_20
11. Stankaitis, P., Iliasov, A., Ait-Ameur, Y., Kobayashi, T., Ishikawa, F., Romanovsky, A.: A refinement based method for developing distributed protocols. In: HASE 2019, pp. 90–97 (2019)
12. Event-B home page. http://www.event-b.org/
13. Zhu, C., Butler, M., Cirstea, C.: Trace semantics and refinement patterns for real-time properties in Event-B models. Sci. Comput. Program. **197** (2020)
14. Sulskus, G., Poppleton, M., Rezazadeh, A.: An interval-based approach to modelling time in event-B. In: Dastani, M., Sirjani, M. (eds.) FSEN 2015. LNCS, vol. 9392, pp. 292–307. Springer, Cham (2015). https://doi.org/10.1007/978-3-319-24644-4_20
15. Rusu, V.: Verifying a sliding-window protocol using PVS. In: Kim, M., Chin, B., Kang, S., Lee, D. (eds.) FORTE 2001. IIFIP, vol. 69, pp. 251–268. Springer, Boston, MA (2002). https://doi.org/10.1007/0-306-47003-9_16
16. Chkliaev, D., Nepomniaschy, V.: Deductive verification of the sliding window protocol. Autom. Control. Comput. Sci. **47**, 12 (2013)
17. Fokkink, W., Pang, J., De Pol, J.: Cones and foci: A mechanical framework for protocol verification. Form. Methods Syst. Des. **29**(1), 1–31 (2006)
18. Van de Snepscheut, J.L.A.: The sliding-window protocol revisited. Formal Aspects Comput. **7**(1), 3–17 (1995)
19. Hoogerwoord, R.R.: A formal derivation of a sliding window protocol. Computer science reports. Technische Universiteit Eindhoven (2006)
20. Méry, D.: Modelling by patterns for correct-by-construction process. In: Margaria, T., Steffen, B. (eds.) ISoLA 2018. LNCS, vol. 11244, pp. 399–423. Springer, Cham (2018). https://doi.org/10.1007/978-3-030-03418-4_24
21. Butler, M.: Incremental design of distributed systems with Event-B. Eng. Methods Tools Softw. Saf. Secur. **22**(131) (2009)
22. Silva, R., Butler, M.: Shared event composition/decomposition in event-B. In: Aichernig, B.K., de Boer, F.S., Bonsangue, M.M. (eds.) FMCO 2010. LNCS, vol. 6957, pp. 122–141. Springer, Heidelberg (2011). https://doi.org/10.1007/978-3-642-25271-6_7

Event-B Formalization of Event-B Contexts

Jean-Paul Bodeveix[1] and Mamoun Filali[2]([✉])

[1] IRIT-UPS, 118 Route de Narbonne, 31062 Toulouse, France
`jean-paul.bodeveix@irit.fr`
[2] IRIT-CNRS, 118 Route de Narbonne, 31062 Toulouse, France
`mamoun.filali@irit.fr`

Abstract. This paper presents an Event-B meta-modelisation of an Event-B project restricted to its context hierarchy which introduces the functional part of a development through sets, constants, axioms and theorems. We study the proposal of a new mechanism for Event-B. It consists in allowing to instantiate in a new context an already proved theorem in a given context. We investigate the validation of the instantiation mechanism in order to prove the validity of imported theorems. We also compare the proposal with similar mechanisms available within some existing theorem provers.

Keywords: Formal methods · Event-B · Meta modelisation

1 Introduction

Event-B [1] is a formal method for the rigorous development of systems. One of its salient features is the Rodin tool [2] which offers an integrated environment for developing and proving. The aim of the EBRP (Enhancing Event-B and Rodin Plus) project[1] is to enhance the framework offered by Rodin in order to better support reuse in Event-B developments thanks to the introduction of generic theories and data types. This enhancement follows the initial work of [4,7]. As a first step of the project, a light extension of the Event-B language and tool has been proposed. In this paper, we investigate an Event-B meta-level description of this extension. An Event-B model consists in a functional model made of an acyclic graph of contexts and a dynamic model using the functional part which consists of successive event-based machine refinements. We focus here on the new reusability mechanism currently studied by the EBRP project for the functional model. It consists in reusing (importing) instances of theorems and axioms considered to be parameterized by the sets and constants declared in their context. The aim of this paper is to validate this importation mechanism: more precisely, we wish to establish the *validity* of an instance of an imported theorem. For this purpose, we propose a meta-level study of Event-B context structure

---

[1] The project EBRP is supported by the French research agency: ANR.

© Springer Nature Switzerland AG 2021
A. Raschke and D. Méry (Eds.): ABZ 2021, LNCS 12709, pp. 66–80, 2021.
https://doi.org/10.1007/978-3-030-77543-8_5

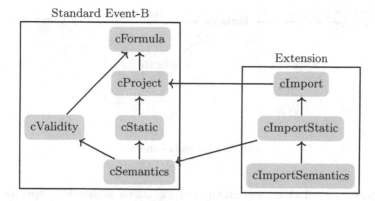

Fig. 1. Metamodel architecture

extended with importation clauses. Figure 1 shows the meta modelisation of a standard Event-B context and its extension with importation/instanciation clauses.

The rest of the paper is structured as follows: Section 2 presents an Event-B meta-model of an Event-B project made of contexts. Section 3 extends this meta-model by the proposed importation mechanism. Section 4 presents similar mechanisms that can be found in other proof environments. Section 5 concludes this paper.

2   Event-B Contexts

In this section, we give a meta-level description of a project made of a hierarchy of contexts. Starting with a high level description of formulas (context `cFormula`), we introduce the subset of valid formulas (context `cValidity`) and describe the project structure as a set of contexts (context `cProject`). Semantics constraints are introduced through the contexts `cStatic` and `cSemantics`. The new importation feature is introduced in the `cImport` context with semantic constraints in `cImportStatic` and `cImportSemantics`. To summarize, the standard representation is structured as a set of contexts corresponding to the left hand side of Fig. 1 and the proposed extension in its right hand side. Moreover, we illustrate the architecture of these contexts through UML-like diagrams[2].

2.1   Formulas

Formulas (see Fig. 2) are modelled at an abstract level. Its free variables are either sets (acting as base types) or constants. A formula denotes either an expression or a predicate. Some expressions (left unspecified here) denote types. The text of a formula is not considered.

---

[2] We could have used the UML-B plugin [9].

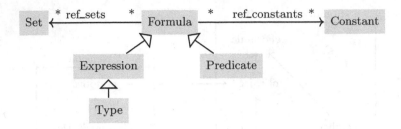

Fig. 2. cFormula context

However, the relations `ref_sets` and `ref_csts` associate respectively the referenced sets and constants. Formulas are partitioned into `Expression` and `Predicate`. Note that we have not modeled the type of an expression and typing constraints. Note also that we could have declared the two relations `ref_sets` and `ref_csts` as functions to power sets of sets or constants. However, it would make more difficult to use the relational composition in order to navigate through the metamodel. These notions are declared through the following labelled (the '@' tag) Event-B axioms[3]:

@Formula_part **partition** (Formula, Expression, Predicate)
   @ref_sets_ty   ref_sets $\in$ Formula $\leftrightarrow$ Set
   @ref_csts_ty   ref_csts $\in$ Formula $\leftrightarrow$ Constant

---

A `Type` is seen as a special expression which only refers to sets (not constants)[4].

@Type_ty Type $\subseteq$ Expression   *// type expressions such as* $\mathbb{P}(A)$ *and* $A \times B$
@ref_type Type $\lhd$ ref_csts $= \emptyset$ *// A type expression doesn't reference a constant*

An important operation on formulas is substitution (`subst`): a set can be replaced by a `Type` and a constant by an `Expression`. The main static property of a substitution: `subst_ref` is that unreferenced sets and constants can be removed from the substitution domain. This expressed through a domain restriction ($\lhd$). We use this property to show that an imported instance of a theorem remains a theorem[5][6][7].

@subst_ty subst $\in$ (Set $\nrightarrow$ Type) $\times$ (Constant $\nrightarrow$ Expression) $\times$ Formula $\rightarrow$ Formula
@subst_ref $\forall$s, c, f·s $\in$ Set$\nrightarrow$Type $\wedge$ c $\in$ Constant$\nrightarrow$Expression $\wedge$ f $\in$ Formula $\Rightarrow$
   subst(s$\mapsto$c$\mapsto$f) = subst(ref_sets [{ f }] $\lhd$ s $\mapsto$ ref_csts [{ f }] $\lhd$ c $\mapsto$ f)

---

[3] $A \leftrightarrow B$ denotes the set of relations from A to B: $A \leftrightarrow B \triangleq \mathcal{P}(A \times B)$.
[4] $\lhd$ denotes domain restriction: $s \lhd r \triangleq r \cap (s \times \mathbf{ran}(r))$.
[5] $x \mapsto y$ denotes the ordered pair (x, y).
[6] $s \nrightarrow t$ denotes a partial function.
[7] $r[s]$ denotes the relational image by r of the set s: $r[s] \triangleq \mathbf{ran}(s \lhd r)$.

2.2   Validity

We first introduce a sequent as a pair formed by a set of hypothesis predicates and a conclusion predicate. Valid sequents are introduced as a subset.

@Sequent_def Sequent = $\mathbb{P}$(Predicate) × Predicate
@Valid_ty Valid ⊆ Sequent

With respect to our concerns, we consider only two axioms about sequents: monotony and substitution of free identifiers which are sets and constants. The first one states that if the hypotheses of a valid sequent are enriched, the sequent remains valid.

@Valid_mono ∀H1,H2,G· H1⊆H2 ∧ H1↦G ∈ Valid ⇒ H2↦G ∈ Valid
@subst_V ∀H,G,s,c· H↦G ∈ Valid ∧ s ∈ Set ⇸ Type ∧ c ∈ Constant ⇸ Expression ⇒
     {h·h ∈ H | subst(s↦c↦h)} ↦ subst(s↦c↦G) ∈ Valid

2.3   Project

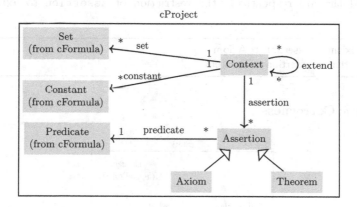

Fig. 3. cProject context

A project contains contexts denoted by the set `Context` linked by the extend relation `extend`. `extend` is declared irreflexive and transitive (Fig. 3).

@extend_ty extend ∈ Context ↔ Context // c1 extends c2
@extend_irr id ∩ extend = ∅
@extend_trans extend;extend ⊆ extend

Sets and constants are defined within contexts. A set or a context is defined once[8]. Note that a set or a constant can be present in unrelated contexts (through the extend relation).

---

[8] $r_1; r_2$ denotes relation composition, usually denoted $r_2 \circ r_1$. It is used to navigate in the metamodel along a chain of links.

@set_ty set ∈ Context ↔ Set
@constant_ty constant ∈ Context ↔ Constant
// a set is defined once in a hierarchy of contexts
@set_uniq_pred (extend; set) ∩ set = ∅
// a constant is defined once in a hierarchy of contexts
@cst_uniq_pred (extend; constant) ∩ constant = ∅

In a context, assertions are stated. An assertion is characterized by a predicate. `assertion` defines the relation between contexts and assertions. Axioms and theorems define distinct assertions.

@assert_ty predicate ∈ Assertion → Predicate
@ass_ty assertion ∈ Context ↔ Assertion
@ass_ctx assertion^{-1} ∈ Assertion → Context
@Assert_fin finite (Assertion)
@Axiom_ty Axiom ⊆ Assertion
@Theorem_ty Theorem ⊆ Assertion
@AxThm Axiom ∩ Theorem = ∅

`axiom` and `thm` are respectively the restriction of `assertion` to axioms and theorems.

@ax_ty axiom = assertion ▷ Axiom
@thm_ty thm = assertion ▷ Theorem

2.4 Static Correcness

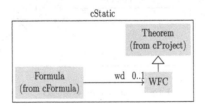

Fig. 4. cStatic context

The static correctness of a project is introduced in the context cStatic. First, sets and constants used by an assertion should be visible from the current context (Fig. 4).

// sets referenced by an assertion are declared
// in the context of the axiom or its ancestry
@sets_ext assertion ; predicate ; ref_sets ; set^{-1} ⊆ id ∪ extend
// sets referenced by an assertion are declared
// in the context of the axiom or its ancestry
@csts_ext assertion ; predicate ; ref_csts ; constant^{-1} ⊆ id ∪ extend

Second, well-formedness conditions are introduced through the subset WFC of Theorem. Static correctness is defined by associating through the wd partial function a wellformedness condition to a formula. It is an assertion to be proven, i.e. a theorem.

```
@WFF_ty WFC ⊆ Theorem
@WD_ty wd ∈ Formula ⤔ WFC
```

Furthermore, a WFC is well formed by construction and thus does not appear in the domain of wd.

```
@WD_WD predicate[WFC] ∩ dom(wd) = ∅
```

The well-definedness condition of a formula does not reference new sets or constants with respect to the initial formula.

```
@sets_wd wd;predicate; ref_sets  ⊆  ref_sets
@csts_wd wd;predicate; ref_csts  ⊆  ref_csts
```

2.5 Semantics

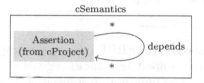

Fig. 5. cSemantics context

The soundness of a context is established through the proof of the validity of its theorems. The proof is abstracted by specifying axioms it uses. We introduce them through the **depends** relation (Fig. 5).

```
@depends_ty depends ∈ Assertion ↔ Assertion
```

An assertion can only depend on assertions that are visible from the current context. We also state that an assertion depends on its well-definedness condition.

```
@depends_extends assertion ;depends ⊆ (id ∪ extend); assertion
@depends_WD predicate;wd ⊆ depends
```

Moreover, the **depends** relation is supposed to be irreflexive and transitive.

```
@depends_irr  id ∩ depends = ∅
@depends_trans depends;depends ⊆ depends
```

A sequent is built from theorems. Its hypotheses are all the assertions on which the theorem depends. Its conclusion is the predicate associated to the theorem itself. The semantics of the **theorem** annotation is thus defined by stating that this sequent is valid.

@THM_V $\forall$t· t $\in$ Theorem $\Rightarrow$ (depends;predicate)[{t}] $\mapsto$ predicate(t) $\in$ Valid

3   Instantiation of Assertions

This section presents a metamodelisation and the validation of an instanciation mechanism proposed by the EBRP project. It is structured as a set of contexts as shown by the right hand side of Fig. 1.

3.1   Informal Presentation

Let us consider a simple generic example with an axiom used to prove a theorem:

```
context gen
sets T
axioms
    @axm1 ∀x,y·x ∈ T∧y ∈ T ⇒ x=y
theorem @th1 T≠∅ ⇒ (∃x·T={x})
end
```

```
context instance1
axioms // import th1 with T mapped to ℤ
    @@th ℤ≠∅ ⇒ (∃x·ℤ={x}) // gen| T:=ℤ|th1
end
```

Within a tool developed by the EBRP project, the proposed syntax to achieve the instanciation of **th1** in context **instance1** is given as a comment. It contains three fields: the context to be imported, instanciation parameters and the name of the target assertion. The instanciated formula can then be automatically generated.

Theorem **th1** can be proved in context gen using axiom **atm1**. **th** is a (considered correct) instance of **th1**. However, while it is expected that importing a theorem should give a theorem, actually **th** is not a theorem. For the assertion "imported theorems are valid" to be valid, a sufficient condition can be that all previous axioms should be imported before as theorems (to be proved), and with the same instance parameters. This is illustrated by the context **instance2** (see Fig. 6).

```
context instance2
axioms // import axm1 and th1
    theorem @@PO ∀x,y·x ∈ ℤ∧y ∈ ℤ ⇒ x=y // gen| T:=ℤ|axm1
    @@th ℤ≠∅ ⇒ (∃x·ℤ={x}) // gen| T:=ℤ|th1
end
```

So imported axioms appearing before imported theorems should become *proof obligations* (thus marked as theorems). Imported theorems should not be proved again and thus appear as axioms.

Here, the theorem PO cannot be proved. It follows that unsoundness of the context `instance2` is clearly pointed out, which is the expected behavior. To sum up, `instance1` should be rejected because an axiom preceding `th1` has not be imported as theorem; `instance2` is accepted by the static type checker but cannot be validated by the user. `instance3` is an example satisfying the static rules and for which the proof obligation for `atm1` instance can be discharged.

```
context instance3 // a correct instance of gen
sets Unit
constants void
axioms
  @part partition(Unit, {void})
  theorem @@atm1 ∀x,y·x ∈ Unit∧y ∈ Unit ⇒ x=y // gen|T:=Unit|axm1
  @@th Unit≠∅ ⇒ (∃x·Unit={x}) // gen|T:=Unit|th1
end
```

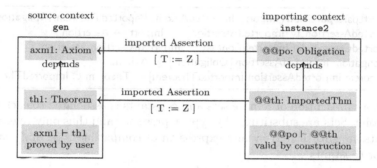

Fig. 6. Imported theorem

3.2 Importation of External Assertions

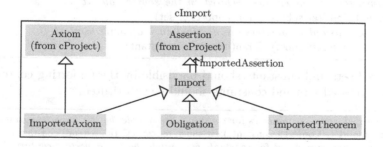

Fig. 7. cImport context

Importation points are added to standard Event-B assertions inside contexts. An importation point is a reference to an assertion of a remote context. Three kind of importations are distinguished: **imported axioms**, **imported theorems** and **obligations** (Fig. 7).

```
axioms
  @Import_ty Import ⊆ Assertion
  @imports_ty imports = assertion ▷ Import
  @obl_ty Obligation ⊆ Import
  @ImportedThm_ty ImportedTheorem ⊆ Import
  @ImportedAxm_ty ImportedAxiom ⊆ Axiom
```

Obligations should be proved by the importing context and are thus declared as a subset of **Theorem**. An **ImportedTheorem** is valid by construction. An **ImportedAxiom** is an instance of an axiom that remains axiomatic in the importing context.

Note that imported theorems should not be proved again. Either their correctness are guaranteed by a meta-level argument (Transformation verification approach), or a proof can be automatically generated and checked (translation validation approach).

```
@Import_part partition (Import, ImportedAxiom, ImportedTheorem, Obligation)
@importedAssertion_ty importedAssertion ∈ Import → Assertion
@importedContext_def importedContext = importedAssertion; assertion⁻¹
@isObligation importedAssertion [ Obligation ] ⊆ Axiom
@isTheorem importedAssertion[ImportedTheorem] ⊆ Theorem ∪ ImportedTheorem
```

Last, a substitution of remote sets and constants is associated to each importation point. Sets are substituted by type expressions and thus only refer to sets. Constants are substituted by any expression of compatible type and may refer to sets or constants[9].

```
@isAxiom importedAssertion[ImportedAxiom] ⊆ Axiom
@importedContext_present importedContext⁻¹;importedAssertion ⊆ axiom
@set_subst_ty set_subst ∈ Import × Set ↠ Type
@cst_subst_ty cst_subst ∈ Import × Constant ↠ Expression
// formal set parameters are declared in the source context
@setp_decl dom(set_subst) ⊆ ran(imports ⊗ set)
// formal constant parameters are declared in the source context
@cstp_decl dom(cst_subst) ⊆ ran(imports ⊗ constant)
```

Referred sets and constants should be visible by the importing context. All remotely accessed sets and constants should be substituted.

```
// constants of actual parameters for constants are visible by the importing context
@cs_rc (imports⊗(Context×Constant));cst_subst; ref_csts ⊆ (id∪extend);constant
// sets of actual parameters for constants are visible by the importing context
@cs_rs (imports⊗(Context×Constant));cst_subst; ref_sets ⊆ (id∪extend);set
// sets of actual parameters for sets are visible by the importing context
@ss_rs (imports⊗(Context×Set));set_subst; ref_sets ⊆ (id∪extend);set
// all sets used by the imported axiom are substituted
@sr_subst importedAssertion; predicate; ref_sets ⊆ dom(set_subst)
```

---

[9] $r \otimes s \triangleq \{x \mapsto (y \mapsto z) \mid x \mapsto y \in r \wedge x \mapsto z \in s\}$.

Note that the model could be refined to introduce typing conditions and constrain expressions to be used by substitutions.

3.3 Static Verification of Importations

In order to guarantee the validity of imported theorems, we link the dependency relation and importation clauses: all dependent assertions on which the imported assertion depends should also be imported. The importing clause should depend on these imports. Derived imports should use compatible substitutions, i.e. common sets or constants should be substituted by the same expressions.

```
@import_depends ∀ctx,imp·
  ctx↦imp ∈ imports ∧ importedAssertion(imp) ∈ Theorem ∪ ImportedTheorem ⇒
    (∀ax· importedAssertion(imp) ↦ ax ∈ depends ⇒
      (∃impa· ctx↦impa ∈ imports ∧
      imp ↦ impa ∈ depends ∧
      ax = importedAssertion(impa) ∧
      (∀s· imp↦s ∈ dom(set_subst) ∧ impa↦s ∈ dom(set_subst) ⇒
        set_subst (imp↦s) = set_subst(impa↦s)) ∧
      (∀c· imp↦c ∈ dom(cst_subst) ∧ impa↦c ∈ dom(cst_subst) ⇒
        cst_subst (imp↦c) = cst_subst(impa↦c))
    ))
```

3.4 Correctness of Theorem Instantiation

We first define the semantics of an imported assertion as the assertion obtained by applying the substitution declared in the importation clause to the imported assertion:

```
@importPredicate ∀imp· imp ∈ Import ⇒
  predicate (imp) = subst({s↦t | (imp↦s)↦t ∈ set_subst} ↦
    {c↦e | (imp↦c)↦e ∈ cst_subst} ↦ predicate(importedAssertion(imp)))
```

The correctness theorem states that the sequent formed by assertions on which the import depends and imported statement is valid. The imported statement can be itself an instance of another distant statement. We thus suppose that the union of the dependency and importation graphs is acyclic. The base case of the result is then given by the following theorem:

```
theorem @ImportValid ∀ctx,imp·ctx↦imp ∈ imports ∧
  importedAssertion(imp) ∈ Theorem ⇒
    (depends; predicate )[{imp}] ↦ predicate (imp) ∈ Valid
```

Given a context `ctx` and an importation point `imp`,

– let `th=importedAssertion(imp)` and suppose it is a theorem. By axiom THM_V of `cSemantics` (Sect. 2.5) the following sequent is valid:

$$(depends;predicate)[\{th\}] \mapsto predicate(th)$$

- Using the set and constant substitutions (S, C) declared in importation point imp, we have predicate(imp) = subst(S,C,predicate(th)) through axiom importPredicate of Sect. 3.4.
- Using axiom subst_V of Sect. 2.2, we can instanciate the sequent to get a new valid sequent Sq_2: subst(S,C)[(depends;predicate)[{th}]] $\mapsto$ th.
- Thanks to axiom import_depends of Sect. 3.3, antecedents of the imported theorem have been imported before with compatible substitutions, i.e. imp depends on these importation points.
- Thanks to axiom subst_ref of Sect. 2.1, applying substitutions (S, C) gives the same result.
- Thus, thanks to the monotonicity of validity (axiom Valid_mono of Sect. 2.2), the sequent concluding on th and containing its dependencies contains enough hypotheses to be valid.

4   Related Concepts

In this section, we review modularity constructs that can be found in various theorem provers. We reuse the same example to illustrate their features and compare them with respect to some key features.

4.1   Section Mechanism in Coq

Variables or hypotheses can be declared in a Coq [10] section and used freely in the rest of the section.

```
Section Gen.
  Variable T: Type.
  Hypothesis axm1: forall (x y: T), x=y.
  Theorem th1: (exists x:T, True) → exists x:T, (forall y:T, x=y).
    intros. destruct H as [x _]. exists x; auto.
  Qed.
End Gen.
```

When the section is closed, variables or hypotheses used by definitions or theorems are made parameters. Here th1 is now seen as a function parameterized by a type T, a proof of axm1 property, and a proof that T is inhabited. An instance of th1 can be obtained through a partial call of th1 with a type and a proof, leading to th definition.

```
Section Instance.
  Inductive unit: Type := One.
  Lemma unit_eq: forall (x y: unit), x=y.
    intros; destruct x; destruct y; auto.
  Qed.
  Definition th := th1 unit unit_eq.
End Instance.
```

Note that it is not necessary to introduce the lemma unit_eq before instantiating the theorem th1: a proof obligation could be generated through the use of Program Definition.

4.2 Module Mechanism in Coq

The theorem th1 is now proved inside a parameterized module (or functor). Its
parameter is typed by the module type tGen declaring T and axm1.

```
Module Type tGen.
  Parameter T: Type.
  Parameter axm1: forall (x y: T), x=y.
End tGen.
Module Gen(U: tGen).
  Theorem th1: (exists x:U.T, True) → exists x:U.T, (forall y, x=y).
    intros. destruct H as [x _]. exists x. apply U.axm1.
  Qed.
End Gen.
```

In order to use the contents of Gen, it must be instanciated by passing a
module compliant with tGen. We introduce the module U defining a one-element
type and proving the required property. Then Gen can be instanciated, which
leads to the module instance I.

```
Module Instance.
  Module U <: tGen.
    Inductive unit: Type := One.  Definition T := unit.
    Lemma axm1: forall (x y: unit), x=y.
      intros; destruct x; destruct y; auto.
    Qed.
  End U.
  Module I := Gen U.
  Definition th := I.th1.
End Instance.
```

4.3 Locales in Isabelle/HOL

Locales [3] introduce a module system in the theorem prover Isabelle [11]. In the
following, the locale gen is parameterized by the variable T typed as a set over
the polymorphic type 'a and states the assumption atm1 over the variables of
the set T. Thanks to this assumption, the theorem th1 is then proved.

```
locale gen =
  fixes T :: "'a set"
  assumes atm1: "∀ x ∈  T. ∀ y ∈  T. x=y"
begin
  theorem th1: shows  "T ≠ ∅ → (∃x ∈ T. T={x})"
  proof using atm1 by blast
  qed
end
```

The constant S is then *defined* as the singleton {1}. The latter set is used
to give an *intepretation* to the locale gen. Then, this intepretation requires to

discharge the assumptions of the locale considered as proof obligations. After unfolding the definition of S and thanks to the powerful tactic `auto` these obligations is automatic.

```
definition  "S = {1}"
interpretation  i: gen "S" unfolding S_def by  unfold_locales  auto
```

4.4   Clones of Why3

In `why3` [5], a theory can declare abstract types and axioms which are used to prove theorems:

```
theory Gen
  type t
  axiom axm1: ∀ x y:t. x=y
  goal th1: (∃ x:t. ⊤ ) → ∃ x:t. ∀ y:t. x=y
end
```

The theory can be instanciated by given values to abstract types. Then axioms automatically become proof obligations. Proof attempts are then performed by the tool, and the contents of the instantiated theory become available to check remaining declarations of the `Instance` theory. The mechanism is less heavy but similar to Coq modules.

```
theory Instance
  type u = Unit
  clone Gen as G with type t = u (* axm1 instance is a proof obligation *)
end
```

4.5   Modules of TLA⁺

In TLA⁺ [8], modules can state assumptions and use them for proofs.

```
1 ———————————————————— MODULE gen ————————————————————
2 CONSTANTS T
3 ASSUME atm1 ≜ ∀x, y ∈ T : x = y
5 THEOREM th1 ≜ T ≠ {} ⟹ (∃x ∈ T : T = {x})
6 ⟨1⟩ QED BY atm1
8 —————————————————————————————————————————————————————
```

In order to instantiate a module, one has to provide the constants (and the variables) that are used in this module. It is also possible to *inherit* the theorems of this instantiated module. However, each such theorem has an additional hypothesis consisting of the assumptions of the instantiated module. Otherwise stated, the reuse of a theorem is possible once the assumptions of its module have been discharged.

```
 1 ┌─────────────────── MODULE instance ───────────────────┐
 3 CONSTANT Z
 5 THEOREM PO ≜ ∀x, y : x ∈ Z ∧ y ∈ Z ⟹ x = y
 7 i ≜ INSTANCE gen WITH T ← Z
 9 THEOREM th ≜ Z ≠ {} ⟹ (∃x ∈ Z : Z = {x})
10 ⟨1⟩ QED BY i!th1, PO
12 └──────────────────────────────────────────────────────┘
```

4.6 Summary

In this section, we summarize some features offered by the module systems of the different proof environments. We have considered three criteria:

- The instantiation syntax: is it possible to extract a single theorem from a module? How formal parameters are given? Do they take implicit values (effective parameter with the same name as the formal parameter)?
- Interaction with provers: is it possible to prove obligations before instantiating the module, during module instantiation, or does the parameter property become a hypothesis of the extracted theorem?
- Extensibility of generic modules: is it possible to extend the generic theory by adding new parameters or new theorems?

	Coq Sections	Coq Modules	Isabelle	Why3	TLA	Rodin
Theorem import	✓					✓
Named parameters				✓	✓	
Implicit parameters					✓	✓
Type param. synthesis	✓	✓	✓	✓	✓	✓
Anticipated proof	✓	✓	✓	✓		✓
Proof obligation	✓		✓	✓		✓
Axioms as assumptions	✓				✓	
Extensibility		✓	✓	✓	✓	✓

As said in the introduction, the current integration of these features is investigated as a light extension of the Rodin tool. Ticks in the Rodin column indicate in-development features. Extensibility is by nature present through Rodin context extension. The other features presented in the table have no impact on the meta-level analysis and will be reflected by the choices done in the final IDE of the tool under construction.

5 Conclusion

In this paper, we have tried to put forward the meta level description of an extension currently studied within the EBRP project. The aim of this meta-modelisation is to validate the expected properties of theorem instantiation. It has been done using current Event-B contexts. This axiomatic formalization could be considered as the formal specification of a static semantic checker of (extended) Event-B models. Also, an interesting evolution of this work could be to take into account the enhancements that are currently being implemented within the EBRP project, as well as some features already available within the plugin theory [7] (inductive types, definition by cases, ...). It would be an interesting validation of these enhancements. A more ambitious aim would be to standardize the syntax and semantics of Event-B [6] through Event-B. It would need to take into account machines, refinements, types, ...

Acknowledgement. We thank the anonymous reviewers for their helpful comments.

References

1. Abrial, J.-R.: Modeling in Event-B: System and Software Engineering. Cambridge University Press, New York (2010)
2. Abrial, J.-R., Butler, M., Hallerstede, S., Hoang, T.S., Mehta, F., Voisin, L.: Rodin: an open toolset for modelling and reasoning in Event-B. Int. J. Softw. Tools Technol. Transf. **12**(6), 447–466 (2010)
3. Ballarin, C.: Locales: a module system for mathematical theories. J. Autom. Reason. **52**(2), 123–153 (2014)
4. Butler, M., Maamria, I.: Practical theory extension in Event-B. In: Liu, Z., Woodcock, J., Zhu, H. (eds.) Theories of Programming and Formal Methods. LNCS, vol. 8051, pp. 67–81. Springer, Heidelberg (2013). https://doi.org/10.1007/978-3-642-39698-4_5
5. Filliâtre, J.-C., Paskevich, A.: Why3—where programs meet provers. In: Felleisen, M., Gardner, P. (eds.) ESOP 2013. LNCS, vol. 7792, pp. 125–128. Springer, Heidelberg (2013). https://doi.org/10.1007/978-3-642-37036-6_8
6. Hallerstede, S.: On the purpose of Event-B proof obligations. Formal Aspects Comput. **23**(1), 133–150 (2011)
7. Hoang, T.S., Voisin, L., Salehi, A., Butler, M.J, Wilkinson, T., Beauger, N.: Theory Plug-in for Rodin 3.x. CoRR, abs/1701.08625 (2017)
8. Lamport, L.: Specifying Systems: The TLA+ Language and Tools for Hardware and Software Engineers. Addison-Wesley, Boston (2002)
9. Snook, C., Butler, M.: UML-B: a plug-in for the Event-B tool set. In: Börger, E., Butler, M., Bowen, J.P., Boca, P. (eds.) ABZ 2008. LNCS, vol. 5238, p. 344. Springer, Heidelberg (2008). https://doi.org/10.1007/978-3-540-87603-8_32
10. The Coq Development Team. The Coq Proof Assistant, January 2021
11. Wenzel, M., Paulson, L.C., Nipkow, T.: The Isabelle framework. In: Mohamed, O.A., Muñoz, C., Tahar, S. (eds.) TPHOLs 2008. LNCS, vol. 5170, pp. 33–38. Springer, Heidelberg (2008). https://doi.org/10.1007/978-3-540-71067-7_7

Validation of Formal Models by Timed Probabilistic Simulation

Fabian Vu[1]([✉])(iD), Michael Leuschel[1]([✉])(iD), and Atif Mashkoor[2]([✉])(iD)

[1] Institut für Informatik, Universität Düsseldorf, Düsseldorf, Germany
{fabian.vu,leuschel}@uni-duesseldorf.de
[2] Institute for Software Systems Engineering, Johannes Kepler University Linz,
Linz, Austria
atif.mashkoor@jku.at

Abstract. The validation of a formal model consists of checking its conformance with actual requirements. In the context of (Event-) B, some temporal aspects can typically be validated by LTL or CTL model checking, while other properties can be validated via interactive animation or trace replay. In this paper, we present a new simulation-based validation technique for (Event-) B models called SimB. The proposed technique uses annotations to construct simulations, taking probabilistic and real-time aspects of the models into account. In this fashion, statistical properties of a single simulation run or a series of runs can be checked (e.g., Monte Carlo estimation or hypothesis tests). SimB complements animation and model checking, and its usability has been assessed via several case studies.

1 Introduction

A typical modeling approach in B [1] and Event-B [2] is to have a generic model for proof, and various instances of the generic model for animation or model checking. The generic model can be *verified* using provers, such as AtelierB[1] or Rodin [3], while the instances can be *verified* or *validated* with ProB [4] using animation and model checking. These two techniques are complementary: proof gives strong guarantees under the assumption of a correct model and can scale to large or infinite-state systems. But it provides limited feedback and typically cannot be used to ensure the presence of a desired real-world behavior. Animation and model checking provide more intuitive user feedback (e.g., in the form of domain-specific visualizations), but typically cannot be applied to generic models and usually cannot be used for exhaustive verification.

---

[1] https://www.atelierb.eu/en/.

This research presented in this paper has been conducted within the IVOIRE project, which is funded by "Deutsche Forschungsgemeinschaft" (DFG) and the Austrian Science Fund (FWF) grant # I 4744-N. The work of Atif Mashkoor has been partly funded by the LIT Secure and Correct Systems Lab sponsored by the province of Upper Austria.

A. Raschke and D. Méry (Eds.): ABZ 2021, LNCS 12709, pp. 81–96, 2021.
https://doi.org/10.1007/978-3-030-77543-8_6

In this paper, we focus on validation, i.e., checking that a formal model is realistic and meets user expectations. Currently, in (Event-) B temporal properties (e.g., liveness) can be validated with LTL model checking, while the presence of features or desired behaviors can be validated via animation and trace replay [5]. However, what is currently missing is the validation via more realistic simulations. In this work, we want to enable validation based on simulation taking into account real-time and probabilistic aspects. Our goal is to develop a lightweight and flexible validation approach, which can also be used for other formalisms (e.g., Z, TLA+, or CSP), and which is capable to accommodate various modelling styles and ways to encode time. Our approach builds on annotations of the respective formal models, processed by a simulator called SIMB built on top of PROB.

As we show later, SIMB can be used for a variety of complimentary validation tasks. Here we sketch one example. Suppose we have a generic Event-B model of a safety-critical component of train movement. This model has an abstraction of the environment, with just the features needed for proofs (e.g., maximal acceleration of other trains). The model may only incorporate a limited, abstract notion of time and may not include information about the likelihood of enabled events. This is where our new technique and tool get involved: we can associate time and probabilities with events of the model, enabling us to conduct realistic simulations as well as to collect statistical information about the formal model, e.g., the likelihood of enabled events or the likelihood of a behavior (within a certain time). Information about timing, probabilities, and interactions between events are not mined from true system executions. So, the challenge for the user is to define simulation parameters in SIMB such that realistic simulations are created. Here it is possible to vary the simulation parameters for the same model to see how it behaves afterwards. This makes it possible to get a better picture of the model's behavior in the real world.

The rest of the paper is organized as follows. Section 2 describes how we encode timing and probabilistic behavior. Its realization in the form of SIMB with the corresponding scheduling algorithm is explained in Sect. 3. Section 4 introduces a class of validation techniques using the presented simulation approach. Section 5 then demonstrates how SIMB is applied to existing examples for validation purposes. In Sect. 6, we compare our approach with published works in the context of simulation and formal methods with probabilistic and timing behaviors. The paper is concluded in Sect. 7.

2   Timed Probabilistic Simulation Principles

This section explains the principles of encoding timing and probabilistic behavior in this work. To make the idea easier to understand, we recall the notion of operations and events in (Event-) B first, which will be referred to as events for the rest of the paper. Events consist of a guard and an action. An event can be fired if it is enabled, i.e., its guard is true. Firing an event executes the corresponding action modifying the current state. Note that events may use parameters and the action may itself be non-deterministic.

(Event-) B is a discrete-time language and events are always executed instantaneously. As shown in Fig. 1, the model switches instantaneously from $c2 = 0$ to $c2 = 2$ without violating the invariant $c2 \in \{0, 2\}$ (i.e., not taking on the intermediate value 1). The (Event-) B method neither provides any guideline about how much time

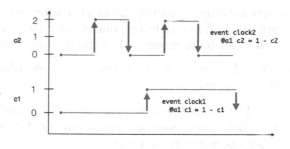

Fig. 1. Clocks example

passes between two event executions[2] nor imposes any restriction on how to choose which enabled event is executed.

```
invariants c1 : {0,1} & c2 : {0,2}
initialisation c1,c2 := 0,0
event clock1 = begin c1 := 1-c1 end
event clock2 = begin c2 := 2-c2 end
```

Listing 1. Examples with Two Independent Clocks

2.1 Encoding Simulation Time

Adding and Adapting Timing Behavior. Our approach works independently of whether time is already part of a model (e.g., in the style as suggested by Rehm et al. [6]) or not. The task of SIMB is to add timing behavior in case time is not already part of the model. Otherwise, SIMB annotations adapt to timing constructs that are part of the model. Furthermore, SIMB simulates many processes in parallel.

Let us first have a look at the small example presented in Listing 1 and Fig. 1. Here, time is not a part of the model. Suppose `clock1` and `clock2` are independent; one ticks every second, the other every 300 ms. Naively, one would increment the simulation time after executing an event. But following this naive approach, it is not possible to model the parallelism of both clocks. However, we can model it if we allow the execution of an event, which triggers other events. So in this example, `clock1` activates itself every second and `clock2` activates itself every 300 ms. This also makes it possible to encode sequential cyclic or acyclic processes like CSP [7].

SIMB also intends to cater for models managing time explicitly, e.g., the models of automotive systems [8]. An important task here is to synchronize SIMB annotations with the model's time, i.e., to adapt the timing behavior. This is enabled by the fact that SIMB activation deadlines do not have to be static but can be computed from the model's constants and variables. Hence,

---

[2] The fact that invariant proof obligations encode an induction proof, the B method implicitly assumes that there are no Zeno runs (i.e., there are no infinitely many events during any given finite time period).

event activation can take an explicit time or deadline variable from the model into account. In conclusion, timing behavior is encoded such that events activate each other to be executed at a specific time in the future.

Event Activation. There are two issues regarding the activation of events with timing behavior:

1. Which event is selected first when many are activated at the same time?
2. How is a new event activation processed if the same event is already queued for execution in the future?

To tackle the first problem, the modeler can assign a priority when annotating an event to control which event is selected first. By default, their priorities are defined by the order they appear in the SIMB annotations. The second problem is solved by adding another event activation to the queue by default, i.e., there are *multiple* activations for the same event. To make our encoding of timing behavior more flexible, it is possible to force SIMB to keep just a single activation. Here, it must be specified whether the new activation should be ignored, or whether the maximum or minimum time of both activations should be taken.

2.2   Encoding Simulation Probabilities

There are three possibilities where probability can be applied to an (Event-) B model:

1. Probabilistic choice in non-deterministic assignments (e.g., x :: S)
2. Probabilistic choice between parameters
3. Probabilistic choice between events

As explained by Hallerstede et al. [9], a model becomes very difficult to understand when it features probabilities besides non-deterministic assignments. Consider different versions to model a coin toss as portrayed in Listing 2. The designer could model it either with non-determinism (1), with a parameter (2), or with two different events (3). In our approach, probabilities are not encoded in the model as we do not extend the B language. Instead, probabilities are encoded in SIMB annotations, which will be explained in Sect. 3 more in detail.

```
toss = BEGIN lastToss :: {{Heads}, {Tails}} END // (1)

toss(c) = PRE c : CoinSide THEN lastToss := {c} END // (2)

toss_heads = BEGIN lastToss := {Heads} END;  // (3)
toss_tails = BEGIN lastToss := {Tails} END
```

Listing 2. Possibilities to Model a Coin Toss

To cover simulation for a wide range of models, it is thus necessary to enable the simulator to control all of the three encodings above. Otherwise, the existing models have to be rewritten to a given format to make the simulator feasible.

3   Simulation Infrastructure

Based on the aforementioned ideas, we now introduce our concept of activating events via annotations combining both timing and probabilistic behavior. An important issue is keeping the syntax and the semantics of the SIMB annotations understandable. Even though both timing and probabilistic behavior are part of SIMB, they should never be mixed up together at the same level. It becomes even more complicated when the modeler has to foresee that an event might be disabled.

After loading a formal model into PROB, corresponding annotations containing probabilistic and timing elements are loaded into the SIMB simulator. Figure 2 shows the architecture of the interaction of SIMB with PROB. SIMB uses PROB to evaluate formulas and to execute events. Again, SIMB manages a scheduling table to store the event activation's scheduled time. An event is executed if

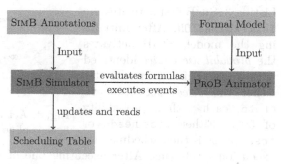

Fig. 2. Interaction of SIMB Simulator with PROB using annotations

these two conditions are met: it is activated for now and it is enabled together with the chosen values for parameters and non-deterministic variables.

Concept of Activation. Initially, SIMB activates SETUP_CONSTANTS and INITIALISATION, whereupon the other events are activated. For each event executed by SimB, the modeler can annotate (multiple) events for activation in the future. There are activations of two kinds:

1. *direct activations* which execute an event after some delay,
2. *probabilistic choices*, which again lead to other activations, each labeled with a probability. The sum of the probabilities must be equal to 1. It is possible to chain multiple such activations together, but eventually, a direct activation must be reached.

Each activation is associated with an id. A *direct activation* always stores the activated event's name. Optionally, it also contains information to control the scheduled time, the (probabilistic) choice between parameters and non-deterministic variables, the priority, additional guards, activation kind, and (multiple) activations to activate other events. In contrast, *probabilistic choice* activations contain the ids of the other activations with the corresponding probabilities.

Those simulation parameters are not mined from true system executions. So, it is the modeler's responsibility to define them such that realistic simulations are created. Using SIMB, the user can vary the simulation parameters for the same model to see how it behaves afterwards. Regarding time and probabilities, it is not only possible to specify constant values, but also to use B formulas which are evaluated in the current state. Thus, it is also possible to vary the simulation parameters within a single simulation.

An example for SIMB annotations for (3) of Listing 2 specifying a coin toss each 500 ms is shown in Listing 3. This results in the corresponding activation diagram graph portrayed in Fig. 3 (*direct activations* in yellow, *probabilistic choice* in red).

When simulating coin tosses with the SIMB annotations in Listing 3, SIMB first activates the INITIALISATION. After initializing the model, SIMB activates the *probabilistic choice* identified with toss. This again activates either the *direct activation* tt or th, each with a probability of 50%. Either toss_heads or toss_tails is then scheduled for

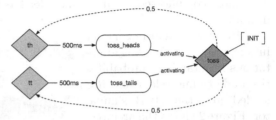

Fig. 3. Activation diagram for Listing 3 (Color figure online)

execution in 500 ms. After executing one of the two events, the *probabilistic choice* toss is triggered again, which results in the next cycle simulating a coin toss.

```
{
  "activations": [
    {"id":"$initialise_machine", "execute":"$initialise_machine",
     "activating":"toss"},
    {"id":"toss", "chooseActivation":{"th": "0.5", "tt": "0.5"}},
    {"id":"th", "execute":"toss_heads", "after":500, "activating":"toss"},
    {"id":"tt", "execute":"toss_tails", "after":500, "activating":"toss"}
  ]
}
```

Listing 3. SIMB Annotations for Coin Toss (3) in Listing 2

Scheduling Algorithm. The scheduling algorithm is separated into two parts: initialization and simulation loop.

Initially, simTime and delay are both assigned to 0. While simTime stores the simulation's current time, delay describes the time for the next scheduled events. The scheduling table st is initialized storing scheduled times for direct activations. They are identified by their id. Again, direct activations are stored in annEvents. In the beginning, the scheduling algorithm activates INITIALISATION and SETUP_CONSTANTS with time(INITIALISATION) = 0 and time(SETUP_CONSTANTS) = 0 respectively. To handle SETUP_CONSTANTS before the INITIALISATION, it is always assigned with a higher priority. For these two special cases, the user is not able to define the priority.

In the following, the simulation loop is described in Algorithm 1. The loop runs until reaching the ending condition, e.g., when the scheduling table is empty and thus no event can be fired anymore, or when the simulator is interrupted by the user. Within each simulation step, simTime is updated. Similarly, each

---

Algorithm 1: Algorithm for Simulation Loop

```
1  procedure simulationLoop()
2    while not endingConditionReached() // Finish at ending condition
3      simTime := simTime + delay // Update time
4      for each annEvent ∈ annEvents // Update scheduling table
5        for each activation ∈ st(id(annEvent))
6          time(activation) := time(activation) - delay
7        executeActivatedEvents() // Execute activated events
8        delay := minimum(activationTimes(st)) // Update delay
9        wait delay // Wait delay (only in real-time simulation)
10 end procedure
```

---

scheduled activation's time is reduced by `delay`. Afterwards, activated events are processed by `executeActivatedEvents`. Finally, `delay` is updated to the time where the next events will be activated. Regarding real-time simulation, i.e., simulation with wall-clock time, this is the waiting time until the next step.

Algorithm 2: Algorithm for Executing Activated Events

```
1  procedure executeActivatedEvents()
2    for each annEvent ∈ annEvents in order of priority

3      // Do not execute if ending condition reached
4      if endingConditionReached()
5        break
6      for each activation ⊂ st(id(annEvent))
7        if time(activation) > 0 // Do not execute if not scheduled
8          break

9        // Remove activation from scheduling table
10       st(id(annEvent)) := st(id(annEvent)) \ {activation}

11       // Select enabled event matching event name,
12       // additional guards, and (probabilistically) chosen
13       // values for parameters and non-deterministic variables
14       transition := selectTransition(activation)

15       if transition exists
16         execute(transition) // Execute transition of activated event
17         activateEvents(annEvent) // activate other events
18 end procedure
```

Now, we refer to Algorithm 2 describing the execution of activated events. To execute scheduled events, the simulator iterates over the direct activations in order of their priority. When no priorities are specified the definition order in the file is used. An activation is scheduled for this step if it is activated now, i.e., its time is equal to 0. It is then removed from the scheduling table. When scheduling activations in the future, each activation is always inserted sorted after the time in the corresponding list. This makes it possible to iterate in each list until an activation is taken, which is not scheduled for this step. Afterwards, an enabled event matching the stored name, additional guards, and the (probabilistic) choice of parameters and non-deterministic variables is selected for execution if it exists. Executing an event activates other events following the *concept of activation*, which is realized by `activateEvents`.

4   Applying SimB for Validation

Real-Time Simulation. Using SimB annotations and the underlying model, a modeler can play a single simulation in real-time, i.e., wall-clock time. This provides a feeling of how the model might behave in practice. The modeler can then manually check whether the model behaves as desired.

Monte Carlo Simulation. SimB also supports Monte Carlo simulations [10]. Here simulations can be performed faster than in real-time (i.e., SimB does not have to wait for the delay to expire, as long as it keeps track of the elapsed time in the model). In SimB, the modeler can specify a start predicate, a start time, or a number of steps that a single generated scenario must have reached. Furthermore, it is possible to define a number of steps, an end predicate, or an end time where the simulation for generating a single scenario should end. In addition to Monte Carlo simulation, the modeler can provide probabilistic and timing properties that are checked on the resulting simulations taking the start condition into account. Two validation techniques are considered here: hypothesis testing [11] and estimation [12]. During Monte Carlo simulation, the simulator also collects statistical information, e.g., the likelihood of enabled events or the likelihood of a behavior (within a certain time).

Given several simulations, a hypothesis, and a significance level, the modeler could ask a question whether to accept or reject the hypothesis. This is done by checking whether a certain property is fulfilled to a given probability. Given several simulations, one could also ask a question about a certain value, which is then estimated. For example, let v_e be the estimated value, and v_d be the desired value, it is then possible to check whether $v_e \in [v_d - \epsilon, v_d + \epsilon]$ for a given ϵ.

Compared to probabilistic (temporal) model checking [13], SimB does not encode a Markov chain which is then used as a state space with probabilities. Furthermore, SimB does not validate probabilistic temporal properties expressed as PCTL [14] formulas. There are also statistical model checking techniques applying Monte Carlo simulation, hypothesis testing, and estimation. Scenarios are generated whereupon PB-LTL formulas [15], or BLTL formulas with a threshold [16] are checked. Since SimB does not check temporal formulas, it is not possible to validate properties over infinite paths. As mentioned before, one can define a start condition as well as an end condition between which a certain property is checked with a probability.

Timed Trace Replay. Given a single simulation run, one can also save it to a trace file with the particular timing encoded as SimB annotations. Afterwards, this timed trace can be re-played in real-time. It does not matter whether the simulation was generated from real-time simulation or Monte Carlo simulation. The resulting SIMB annotations do not contain any probabilistic elements. Consider a simulation generating a trace with length n where the timestamp of the i-th event is $ts(i)$. It is then possible to generate SIMB annotations where executing event i activates the event $i+1$ with annotation $i+1$ and with a scheduled time of $ts(i+1) - ts(i)$. Nevertheless, it is still challenging to check whether a timed trace can be re-generated from a modified model or SIMB annotation. It might then be necessary to save more information about the simulator's history, e.g., which probabilistic choices have been taken into account or which activations have been discarded.

5   Case Studies

This section demonstrates the application of SIMB to various case studies. See Table 1 for a complete list of applied case studies, which are accessible online[3].

Real-Time Simulation and Timed Trace Replay. In the context of real-time simulation, we only focus on the automotive case study [8]. As a case study, the driver's inputs on the pitman controller and the warning lights are simulated. Every time the driver activates the pitman controller or the warning lights, a sequence of events blinking the lights until the driver's next input is triggered.

A property to be validated is, e.g., that the corresponding lights are turned on with full intensity within a certain time after the driver makes an input on the pitman controller. Another property for validation is that the lights never turn on until the driver makes an input on the pitman controller or the warning lights button.

Within the model, the time is modeled as a variable that is increased by events, which are responsible for passing the time, passing the time until the next deadline, blinking the lights and passing the time until the next deadline, and passing the time until the next deadline with a timeout. By following the principles of our simulation approach (see Sect. 2), it was possible to adapt to the model's timing specification.

Figure 4 shows an actual simulation illustrated as VISB [17] visualizations. Using VISB alone, one can create SVG images and an associated VISB file for a model. Within the VISB file, it is possible to define which operation is triggered when clicking an SVG element. Here, one can also manipulate the style of the images by using B formulas which are evaluated in each state. In combination with SIMB, a simulation can then be seen as an animated picture similar to a GIF picture. As shown in Fig. 4a, the engine is turned on, the pitman controller is in a neutral position, the warning lights button is not pressed, and the lights are turned off. After 1.7 s have passed, the driver decides to move the pitman controller to `Upward7`, which activates the lights on the right-hand side (see Fig. 4b). With a delay of 100 ms, the lights on the right-hand side turn on whereupon they blink every 500 ms (see Fig. 4c–Fig. 4e).

---

[3] Available at https://github.com/favu100/SimB-examples.

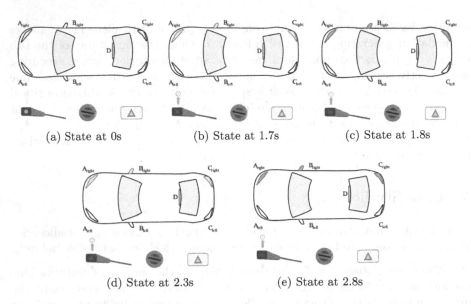

(a) State at 0s   (b) State at 1.7s   (c) State at 1.8s

(d) State at 2.3s   (e) State at 2.8s

Fig. 4. Simulation example for the automotive case study [8]

Timed traces are successfully captured from the real-time simulation as well as Monte Carlo simulation. As explained before, they are stored as SimB annotations. Thus, re-playing them works similar to real-time simulation.

Monte Carlo Simulation. SimB is also used for Monte Carlo simulation, together with hypothesis testing and estimation, to validate the timing and probabilistic behavior of formal models. The results are shown in Table 1. Validated properties also include "almost-certain" properties, i.e., properties describing a random event to occur with probability 1. The examples Dueling Cowboys, Tourists (aka Rabin's Choice Coordination), and Leader Election (aka Herman's probabilistic self-stabilization) are taken from the work of Hallerstede et al. [9] and Hoang [18]. All experiments are done with a fixed seed[4] in ProB2-UI[5] on a MacBook Air with 8 GB of RAM and a 1.6 GHz Intel i5 processor with two cores. A significance level and an ϵ-value are set for hypothesis testing and estimation respectively (as described in Sect. 4). Both values are set to 1% for 100 runs. Again, they are set to 0.1% for $\geq 10\,000$ runs.

Simulations with $\geq 10\,000$ runs were calculated within 4 min, with each of them executing more than 500 000 events. In contrast, those simulations with only 100 runs take up to 1 h to terminate. Here, a significantly lower number of events ($\leq 65\,000$ events) are executed for each simulation. Currently, SimB evaluates all outgoing transitions before choosing one. This can obviously lead to performance issues. Particularly, a very large number of transitions are evaluated

---

[4] Seed is a number used to initialize the random number generator to make the results are reproducible. We used 1000 as seed.

[5] https://github.com/hhu-stups/prob2_ui.

in the simulations with 100 runs. For the Dueling Cowboys we produced a more abstract version with a smaller state space, enabling us to simulate 10 000 runs in less than 13 s. In future, SIMB could be improved such that it only evaluates a single transition given the probabilistic annotations.

Table 1. Application of SIMB validation techniques based on Monte Carlo simulation to case studies with number of runs, number of Evaluated Transitions (ET), Runtime in Seconds (RT), and the result of validation

Model	Simulation	Property	Runs	ET	RT	Result
Coin toss	Fair Tosses	Heads in 50% of all Tosses	1 000 000	7	8.19	✓ (49.93%)
		Eventually Heads in 100 Tosses	10 000	7	3.43	✓ (100%)
Rolling dice	Fair Dices	6 in 16.67% of all Rolls	1 000 000	43	10.33	✓ (16.66%)
		Eventually 6 in 100 Rolls	10 000	43	6.09	✓ (100%)
Dueling cowboys	100 Cowboys,	Termination in 125 Shots	100	1 720 854	1676.06	✗ (56%)
	80% Accuracy	Termination in 250 Shots	100	1 723 302	1703.74	✓ (100%)
Dueling cowboys (abstract)	100 Cowboys,	Termination in 125 Shots	10 000	201	11.01	✗ (63.13%)
	80% Accuracy	Termination in 250 Shots	10 000	201	12.51	✓ (100%)
Tourists	100 Tourists	Termination in 125 Moves	100	956 468	2019.1	✗ (0%)
		Termination in 300 Moves	100	1 081 099	3195.14	✓ (100%)
Leader Election	10 Nodes	Termination in 250 Steps	10 000	37 917	201.6	✗ (99.46%)
		Termination in 500 Steps	10 000	37 884	201.36	✓ (100%)
Traffic Light (TL)	Cars TL from	Red to Green in 0.5 s for Cars	1 000 000	5	9.61	✗ (0%)
	Red to Green	Red to Green in 1 s for Cars	1 000 000	5	9.84	✓ (100%)
Lift	Basement to 2nd floor	Reaching 2nd floor in 10 s	1 000 000	47	48.11	✗ (0%)
		Reaching 2nd floor in 20 s	1 000 000	47	46.57	✓ (100%)
Lift	Basement to stop at 1st floor stop at 1st floor	Reaching 2nd floor in 20 s	1 000 000	70	78.36	✗ (0%)
Automotive Case Study[a]	Random Input on Pitman Controller and Hazard Warning Signal with Engine on	Left light blinks 100 ms with full intensity after moving pitman to Downward7	10 000	106	22.73	✗ (99.17%)
		Left light blinks 500 ms with full intensity after moving pitman to Downward7	10 000	106	22.37	✓ (100%)
		Lights never turn on until it is activated via pitman or warning lights	10 000	74	9.51	✓ (100%)

[a] Specification was optimized for model checking

6   Related Work

In this section, we compare our work with the state of the art in the field of modeling and simulation of probabilistic and timing behaviors.

Modeling and Verification of Probabilistic Behavior. Hallerstede et al. [9] introduce probabilistic events for Event-B in which the modeler can use probabilistic assignments in place of non-deterministic assignments (but not probabilistic choice between events nor for parameters, and there are no explicit values for the probabilities used). Based on this work, Hoang [18] presents an approach to verify almost-certain properties. In contrast, SIMB simulates existing models by using additional annotations. This makes it possible to gain better insights on how the model would behave in a real-world application. SIMB can use statistical techniques to validate the presence of a desired behavior with detailed feedback. We achieve this by building on top of the semantics of (Event-)B, not by changing it at the core. Our approach is more empirical than formal and proof-based.

Legay et al. [13,16] provide an overview of statistical model checking including probabilistic model checking and numerical approaches. While the former is applied to a Markov chain (used as a state space with probabilities), the latter approximates certain values during validation. We do not encode a Markov chain in SIMB and our work does not apply probabilistic model checking. Furthermore, SIMB does not validate probabilistic temporal properties expressed as PCTL [14] formulas. There are also statistical model checking techniques applying Monte Carlo simulation, hypothesis testing, and estimation. This is done by generating simulations and checking timing properties expressed as BLTL formulas with a threshold. Abdellatif et al. [15] present a simulation-based approach to generate attacking scenarios to validate probabilistic properties expressed as PB-LTL formulas in a model of smart contracts and the blockchain. Similar to our work, safety properties are also checked with fault tolerance and estimation of error probability. In contrast, we provide a property that is checked for each generated simulation, e.g., whether a predicate is eventually true between the starting condition and the ending condition of a simulation. Since SIMB does not check temporal formulas, it is not possible to validate probabilistic properties over infinite paths.

Modeling and Verification of Timing Behavior. To model and verify real-time behavior, the modeler could use formalisms, such as timed automaton [19], with existing model checkers, e.g., Uppaal [20]. There are also approaches to verify both probabilistic and real-time behavior, e.g., by using the model checker PRISM which is applied to probabilistic timed automaton [21]. To check such properties in (probabilistic) timed automata, *reachability analysis* is applied. Its task is to check whether a state is reachable with the given timing (and probabilistic) properties. Our proposed approach is a lightweight solution to simulate existing models to gain additional insights into how they might behave in practice. SIMB simulates a model until a certain condition, a certain time, or a certain number of steps is reached. Probabilistic and timing properties are then validated with statistical methods on the resulting traces without applying *reachability analysis*. Thus, SIMB is not meant to replace other approaches based on timed automata. Abdellatif et al. [22] present a scheduling approach to check whether the modeled program is implementable, holding the defined timing properties. Again, one can also model concrete time in discrete-time formalisms, e.g., by following an approach presented by Leslie Lamport for TLA+ [23], or Event-B by following a timing constraint pattern discussed by Rehm et al. [6] and Mashkoor et al. [5]. Timing properties can then be verified with existing provers and model checkers in the corresponding languages. As aforementioned, our work simulates the underlying (Event-) B model by using annotations for timing and probabilistic behavior. SIMB annotations can be used to match the modeled time, see, e.g., the automotive case study [8]. Furthermore, SIMB can also simulate user behavior.

Simulators. JEB [24] is a framework, which translates Event-B models into JavaScript programs for simulation. Models are sometimes too abstract for animation tools such as PROB. To enable validation of these models anyway without refining, they are translated into executable programs. One can also insert pieces of code to control the simulation. The challenge is to define the *fidelity* property between the model and its translation [25]. In contrast, the task of SIMB is not to make models executable, but to simulate executable models to apply statistical validation techniques. Therefore, SIMB is built on top of the PROB animator.

Similar to our approach, Dieumegard et al. [26] present a simulator for an anti-collision function of a small robotic rover based on Event-B to understand how the specification behaves. Note that the simulator presented by Dieumegard et al. is not a generic one. So, it is limited to the robotic rover case study.

Co-Simulation. In VDM, simulation is already more common than animation or model checking. It has now been extended by a co-simulation toolset named INTO-CPS [27]. INTO-CPS tools also contain design space exploration implemented with search algorithms where simulation parameters may vary and scenarios outcomes evaluated. Thus, INTO-CPS can search for optimal simulation parameters. In SIMB, it is still a challenge for the modeler to choose simulation parameters such that realistic scenarios are generated. The modeler could, e.g., vary the simulation parameters for the same model to see how it behaves afterwards. But this process has to be done manually. Compared to this co-simulation tool-set, our approach is somewhat limited but much more lightweight. For example, there is neither a continuous simulation tool running nor an FMI interface in SIMB. In the future, SIMB annotations could actually be used on top of a co-simulation using (Event-) B.

Other Formalisms. It would have been possible to use CSP control annotations for (Event-) B models [28], as available in PROB [29]. But, CSP does not cater to probabilities or time. The Timed CSP interpreter from [30] is not available in the current release of PROB, and also lacks probabilistic features. There are also formalisms combining probabilistic and timing behavior such as Probabilistic Time Petri Nets [31]. Since this formalism is not supported in PROB, it would be necessary to implement an interpreter to control the model. SIMB is designed to be as simple as possible, but strong enough for simulation of models with probabilistic and timing behavior. Moreover, on a technical side, our annotations also work for other formalisms (such as TLA+).

7 Conclusion and Future Work

In this paper, we presented SIMB – a simulator for formal models, which adapts the concept of activations annotating events with timing and probabilistic elements. Here, it was particularly important to separate probabilistic and timing behavior from each other to keep the syntax and semantics of SIMB understandable. By building SIMB on top of PROB, it was possible to support formalisms that are supported by PROB such as B, Event-B, Z, TLA+, and CSP. SIMB is capable of simulating environment inputs, e.g., by users, and models' behaviors.

In this work, the usability of SIMB was demonstrated in several examples. SIMB can either be used to extend existing models by timing and probabilistic behavior, or adapt to models where time is modeled as a variable. SIMB is capable to validate formal models using Monte Carlo simulation, hypothesis testing, estimation, and timed trace replay. Using Monte Carlo simulation, the modeler can generate scenarios to gain insights into how the model might behave in real-world. It is then possible to replay them with timing behavior, or to validate timing and probabilistic properties with hypothesis testing and estimation.

More information on SIMB with screenshots and a tutorial is available at:

https://prob.hhu.de/w/index.php?title=SimB

As future work, it would be possible to add more statistical validation methods. Furthermore, the performance of SIMB could be improved. On the one hand, SIMB should compute a single transition given the defined probabilistic annotations, rather than computing all and choosing afterwards. On the other hand, one could apply code generation for SIMB. To cover a wide range of models, it would be necessary to target generated code from other high-level code generators such as B2PROGRAM [32], EventB2Java [33] or Asm2C++ [34]. Additionally, we would also like to describe the semantics of SIMB in formal logic. One could then implement an interpreter and integrate it into PROB's core. This could be a way to reduce the overhead of SIMB to PROB. Furthermore, it is then possible to animate and model check (Event-) B models together with SIMB annotations.

Regarding the future, one could investigate how SIMB can be used for co-simulation. Furthermore, it is still a challenge for the modeler to choose simulation parameters such that realistic scenarios are generated. So, another future work would be to analyze how optimal simulation parameters could be explored.

Eventually, we intend to use SIMB in the context of *validation obligations* [35], which is the idea of breaking down the validation of a formal model into smaller tasks and associating them with each refinement step. Validations should then be applicable and re-usable for the whole software development life cycle.

References

1. Abrial, J.-R.: The B-Book. Cambridge University Press, Cambridge (1996)
2. Abrial, J.-R.: Modeling in Event-B: System and Software Engineering. Cambridge University Press, Cambridge (2010)
3. Abrial, J.-R., Butler, M., Hallerstede, S., Hoang, T.S., Mehta, F., Voisin, L.: Rodin: an open toolset for modelling and reasoning in Event-B. Int. J. Softw. Tools Technol. Transf. **12**(6), 447–466 (2010)
4. Leuschel, M., Butler, M.: ProB: an automated analysis toolset for the B method. STTT **10**(2), 185–203 (2008)
5. Mashkoor, A., Jacquot, J.-P.: Utilizing Event-B for domain engineering: a critical analysis. Requir. Eng. **16**(3), 191–207 (2011)
6. Rehm, J., Cansell, D.: Proved development of the real-time properties of the IEEE 1394 root contention protocol with the event-B method. In: Proccedings ISoLA, pp. 179–190 (2007)
7. Hoare, T.: Communicating sequential processes. Commun. ACM **21**(8), 666–677 (1978)
8. Leuschel, M., Mutz, M., Werth, M.: Modelling and validating an automotive system in classical B and Event-B. In: Raschke, A., Méry, D., Houdek, F. (eds.) ABZ 2020. LNCS, vol. 12071, pp. 335–350. Springer, Cham (2020). https://doi.org/10.1007/978-3-030-48077-6_27
9. Hallerstede, S., Hoang, T.S.: Qualitative probabilistic modelling in Event-B. In: Davies, J., Gibbons, J. (eds.) IFM 2007. LNCS, vol. 4591, pp. 293–312. Springer, Heidelberg (2007). https://doi.org/10.1007/978-3-540-73210-5_16
10. Mooney, C.Z.: Monte Carlo Simulation, vol. 116, Sage Publications (1997)
11. Kendall, M.G., Stuart, A., Keith Ord, J.: Kendall's Advanced Theory of Statistics. Oxford University Press, Oxford (1987)
12. Fisher, R.A.: Theory of statistical estimation. Math. Proc. Cambridge Philos. Soc. **22**(5), 700–725 (1925)
13. Legay, A., Delahaye, B., Bensalem, S.: Statistical model checking: an overview. In: Barringer, H., et al. (eds.) RV 2010. LNCS, vol. 6418, pp. 122–135. Springer, Heidelberg (2010). https://doi.org/10.1007/978-3-642-16612-9_11
14. Hansson, H., Jonsson, B.: A logic for reasoning about time and reliability. Formal Aspects Comput. **6**, 512–535 (1995)
15. Abdellatif, T., Brousmiche, K.-L.: Formal verification of smart contracts based on users and blockchain behaviors models. In: Proceedings NTMS, pp. 1–5 (2018)
16. Legay, A., Lukina, A., Traonouez, L.M., Yang, J., Smolka, S.A., Grosu, R.: Statistical model checking. In: Steffen, B., Woeginger, G. (eds.) Computing and Software Science. LNCS, vol. 10000, pp. 478–504. Springer, Cham (2019). https://doi.org/10.1007/978-3-319-91908-9_23
17. Werth, M., Leuschel, M.: VisB: a lightweight tool to visualize formal models with SVG graphics. In: Raschke, A., Méry, D., Houdek, F. (eds.) ABZ 2020. LNCS, vol. 12071, pp. 260–265. Springer, Cham (2020). https://doi.org/10.1007/978-3-030-48077-6_21
18. Hoang, T.S.: Reasoning about almost-certain convergence properties using Event-B. In: Proceedings AVoCS. LNCS, vol. 81, pp. 108–121 (2014)
19. Alur, R., Dill, D.L.: A theory of timed automata. Theoret. Comput. Sci. **126**, 183–235 (1994)

20. Bengtsson, J., Larsen, K., Larsson, F., Pettersson, P., Yi, W.: UPPAAL—a tool suite for automatic verification of real-time systems. In: Alur, R., Henzinger, T.A., Sontag, E.D. (eds.) HS 1995. LNCS, vol. 1066, pp. 232–243. Springer, Heidelberg (1996). https://doi.org/10.1007/BFb0020949

21. Kwiatkowska, M., Norman, G., Sproston, J., Wang, F.: Symbolic model checking for probabilistic timed automata. In: Lakhnech, Y., Yovine, S. (eds.) FORMATS/FTRTFT -2004. LNCS, vol. 3253, pp. 293–308. Springer, Heidelberg (2004). https://doi.org/10.1007/978-3-540-30206-3_21

22. Abdellatif, T., Combaz, J., Sifakis, J.: Model-based implementation of real-time applications. In: Proceedings of the Tenth ACM International Conference on Embedded Software, pp. 229–238. ACM (2010)

23. Lamport, L.: Real-time model checking is really simple. In: Borrione, D., Paul, W. (eds.) CHARME 2005. LNCS, vol. 3725, pp. 162–175. Springer, Heidelberg (2005). https://doi.org/10.1007/11560548_14

24. Mashkoor, A., Yang, F., Jacquot, J.-P.: Refinement-based Validation of Event-B Specifications. Softw. Syst. Model. **16**(3), 789–808 (2016). https://doi.org/10.1007/s10270-016-0514-4

25. Mashkoor, A., Jacquot, J.-P.: Validation of formal specifications through transformation and animation. Requirements Eng. **22**(4), 433–451 (2016). https://doi.org/10.1007/s00766-016-0246-6

26. Dieumegard, A., Ge, N., Jenn, E.: Event-B at work: some lessons learnt from an application to a robot anti-collision function. In: Barrett, C., Davies, M., Kahsai, T. (eds.) NFM 2017. LNCS, vol. 10227, pp. 327–341. Springer, Cham (2017). https://doi.org/10.1007/978-3-319-57288-8_24

27. Thule, C., Lausdahl, K., Gomes, C., Meisl, G., Larsen, P.G.: Maestro: the INTO-CPS co-simulation framework. Simul. Model. Pract. Theory **92**, 45–61 (2019)

28. Ifill, W., Schneider, S., Treharne, H.: Augmenting B with control annotations. In: Julliand, J., Kouchnarenko, O. (eds.) B 2007. LNCS, vol. 4355, pp. 34–48. Springer, Heidelberg (2006). https://doi.org/10.1007/11955757_6

29. Butler, M., Leuschel, M.: Combining CSP and B for specification and property verification. In: Fitzgerald, J., Hayes, I.J., Tarlecki, A. (eds.) FM 2005. LNCS, vol. 3582, pp. 221–236. Springer, Heidelberg (2005). https://doi.org/10.1007/11526841_16

30. Dragon, M., Gimblett, A., Roggenbach, M.: A simulator for timed CSP. In: Proceedings AVoCS. Electronic Communications of the EASST, vol. 46 (2011)

31. Emzivat, Y., Delahaye, B., Lime, D., Roux, O.H.: Probabilistic time petri nets. In: Kordon, F., Moldt, D. (eds.) PETRI NETS 2016. LNCS, vol. 9698, pp. 261–280. Springer, Cham (2016). https://doi.org/10.1007/978-3-319-39086-4_16

32. Vu, F., Hansen, D., Körner, P., Leuschel, M.: A multi-target code generator for high-level B. In: Proceedings iFM 2019, pp. 456–473 (2019)

33. Cataño, N., Rivera, V.: EventB2Java: a code generator for Event-B. In: Rayadurgam, S., Tkachuk, O. (eds.) NFM 2016. LNCS, vol. 9690, pp. 166–171. Springer, Cham (2016). https://doi.org/10.1007/978-3-319-40648-0_13

34. Bonfanti, S., Gargantini, A., Mashkoor, A.: Design and validation of a C++ code generator from abstract state machines specifications. J. Softw. Evol. Process. **32**(2), e2205 (2020)

35. Mashkoor, A., Leuschel, M., Egyed, A.: Validation obligations: a novel approach to check compliance between requirements and their formal specification. In: ICSE 2021 NIER (2021)

Short Articles

Sterling: A Web-Based Visualizer for Relational Modeling Languages

Tristan Dyer[1(✉)] and John Baugh[2]

[1] Brown University, Providence, RI, USA
tristan_dyer@brown.edu
[2] North Carolina State University, Raleigh, NC, USA

Abstract. We introduce Sterling, a web-based visualization tool that provides interactive views of relational models and allows users to create custom visualizations using modern JavaScript libraries like D3 and Cytoscape. We outline its design goals and architecture, and describe custom visualizations developed with Sterling that enable verification studies of scientific software used in production. While development is driven primarily by the Alloy community, other relational modeling languages are accommodated by Sterling's data agnostic architecture.

Keywords: Alloy · Sterling · Formal methods · Visualization

1 Introduction

Model finding tools like Alloy enable a lightweight approach to design and reasoning about complex software systems. Such tools provide push-button analysis for both checking assertions within bounded scopes, and for generating instances that satisfy a property of interest. An attractive feature of Alloy is the immediate feedback provided by visualizations, allowing users to inspect instances and counterexamples in order to identify design problems. The ability to communicate visual information *intuitively* therefore plays a key role in determining the effectiveness of interactions with the user [5].

The built-in visualizer in the Alloy Analyzer can display an instance as a directed graph in which nodes represent atoms and edges represent tuples of relations. To help users better understand an instance, basic properties of the graph such as color and shape can be customized, and graph nodes can be repositioned manually to achieve a clearer layout. Additionally, the graph view supports "projection," a feature most commonly used with models of dynamic systems, in which an instance is displayed from the perspective of an atom or set of atoms. When an instance of such a model is projected over time, the user is able to step through snapshots of individual states in sequence.

Despite these capabilities, some instances can be difficult to interpret as models grow in size and complexity. Some well-known issues, for instance, include the inability to drag nodes out of the rows into which they are initially laid out [3,8].

A. Raschke and D. Méry (Eds.): ABZ 2021, LNCS 12709, pp. 99–104, 2021.
https://doi.org/10.1007/978-3-030-77543-8_7

In addition, the graph layout is recalculated any time a new instance is generated or the projection is changed, so the user is forced to reinterpret the entire graph if, for example, they are stepping through the state atoms in sequence [3,9,12]. Various approaches have been proposed to address these and other issues, either by extending the existing visualizer [12] or by introducing new tools [3,5,8]. Our own experiences with Alloy in the field of scientific computing have highlighted the need for better visualization approaches in general, and for an interface that can also depict *spatial* relationships—not just topological ones—while maintaining *consistency* in those relationships when dynamic updates occur, as they do in problems with time-varying state.

For instance, in one such study, Baugh and Altuntas [2] use Alloy to explore implementation choices and ensure soundness of an extension made to a large-scale storm surge application used in production. To be physically meaningful, models representing finite element meshes—which can be thought of as a triangulation of a surface—are constrained to include only those that have a planar embedding, and therefore do not contain overlapping triangles. Working with relational depictions alone means "untangling" each instance as it occurs, leading to the study's conclusion that "more than any extension to Alloy, what would have benefited our study most is a tool capable of automatically producing planar embeddings of meshes from Alloy instances, which proved to be tedious to do by hand."

A subsequent study [4] with Alloy focused on bounded verification of sparse matrix formats, which use array indirection and other structure to avoid storing zeroes. Dense matrices are modeled as relations mapping indices to values, producing dozens of tuples that clutter and overrun any visualization attempt with edges. The sparse matrices themselves, and the dynamic state changes that accompany them for operations like matrix multiplication, make visualizations nearly impossible to interpret.

2   Sterling Design and Architecture

Motivated by these studies, and drawing on feedback and suggestions from the 2018 Workshop on the Future of Alloy, we have developed an approach that builds on the strengths of existing visualizations. Sterling's design is based on the following principles: the visualizer should (1) implement and extend the capabilities already present in Alloy, (2) employ a modern architecture built using popular languages and well established libraries, and (3) provide functionality for creating domain specific visualizations.

Consistent with these principles, Sterling is a web application,[1] built upon a popular web technology stack using the React and Redux libraries, and packaged with a custom build of Alloy. A client-server relationship is established between Sterling and Alloy by an embedded web server, enabling instances to be immediately visualized in Sterling as they are generated by Alloy. The user interface is similar to Alloy's own, providing graph and table views which extend the

---

[1] A Sterling demo with examples can be found at https://sterling-js.github.io.

functionality of their counterparts in Alloy, while adding a "script" view that provides users with the ability to create custom visualizations from instance data by writing JavaScript code. Communication between Alloy and the individual views is managed using a mediated model-view architecture, illustrated in Fig. 1. Consequently, other relational logic and model finding tools may also employ Sterling for visualization, so long as data is provided to Sterling in the Alloy XML format.

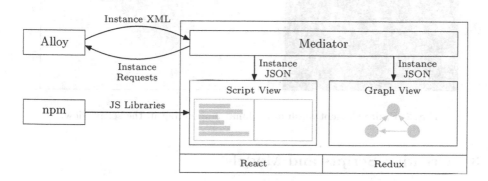

Fig. 1. The sterling architecture.

The Sterling graph view offers the same functionality as the Alloy graph view, but also provides a few key extensions. Most notably, graph elements are not restricted to rows, and users may freely arrange graphical elements to make the display more readable. Furthermore, the layout algorithm is not automatically executed when the projection is changed or a new instance is generated, and so graphical elements remain static as users step through stateful models and generate instances.

The Sterling script view provides an environment for the rapid development of custom visualizations by bringing together a text editor, a blank canvas, and a JavaScript execution environment, giving users a basic "code sandbox" in which they can create visualizations based on instance data using their favorite JS libraries. Within the script view environment, all instance data—the signatures, fields, atoms, and tuples—are exposed as JS variables. Additionally, users have direct access to the npm package repository, which can be used to add visualization (or any other useful) libraries to the scripting environment. This combination enables, for example, a user to bind atoms to shapes using the D3 visualization library, and to calculate their positions based on relationships defined by tuples. We have found this paradigm to be particularly useful for visualizing instances of models with inherent spatial properties. For example, a planar embedding of a finite element mesh, as previously described, is shown in Fig. 2. More custom visualizations, including ones for sparse matrices and some common puzzles, can be found in the interactive demo on the Sterling website.

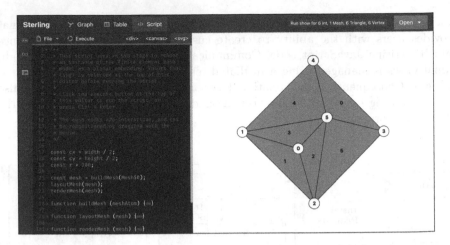

Fig. 2. Finite element mesh as a planar embedding in the script view.

3   Creating Scripts and Models

The script view is designed to integrate with the iterative design process that is typical of Alloy, and as such, users receive the same kind of immediate visual feedback provided by the graph view, with the added benefit of complete control over the visual approach used to display instances. In typical usage of the script view, a user begins by writing a model and executing a command to generate an instance. The instance is automatically sent to Sterling, where the user then writes a script in the script view. To execute the script and generate the visualization, the user presses "Ctrl+Enter" or clicks the "Execute" button located at the top of the script editor. The user can continue to refine the visualization by editing and rerunning the script, or use the "Next" button to explore more instances. Each time an instance is generated, Sterling automatically executes the script to re-render the visualization. This automatic execution continues when the user returns to Alloy, refines the model, and generates new instances. If the model is changed in a way that causes the visualization script to throw an error, the user is notified, and they must then update the visualization script to reflect the new model.

In practice we have found visualization scripts typically start out simple and grow in complexity alongside the model. For example, early iterations of the previously described matrix models employed the Cytoscape JS library to create interactive "snapshot" views of instances, as shown in Fig. 3a. These snapshots, described by Jackson [6], proved useful in the development and understanding of both the hierarchical and relational structure of the models. As the structure of the models became more concrete and focus shifted to modeling the behavior of sparse matrix operations, the visualization script evolved to provide a more realistic view of matrices as well as a clear depiction of state change, as shown in Fig. 3b.

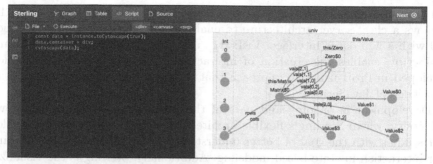

(a) A snapshot view of a dense matrix

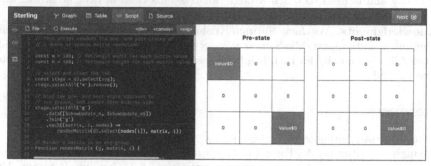

(b) A sparse matrix update operation

Fig. 3. Scripts for matrix models at (a) early and (b) late stages of development.

For users who are comfortable using JavaScript, particularly those with experience using popular JS visualization libraries, creating custom visualizations from Alloy instances is straightforward. To support users with little or no experience, the Sterling website provides tutorials and numerous examples that demonstrate basic visualization techniques. Furthermore, scripts capable of generating custom visualizations for some common modeling paradigms, such as binary trees and directed graphs, are available on the website and can be used out-of-the-box.

4   Conclusion

Sterling addresses some of the common issues identified with existing visualizations in Alloy, and introduces a script view to enable development of custom visualizations without sacrificing the immediate visual feedback provided by the Alloy Analyzer. The Sterling architecture and visualization approach take inspiration from other tools developed to address certain visualization challenges in Alloy and other formalisms. Alloy4Fun [8] and BMotionWeb [7] are both web-based tools that leverage the popularity of the JavaScript programming language and the availability of robust data visualization libraries, and PVSio-Web [10] employs a client-server architecture to enable coupling of a formal verification tool with a web-based interface. VisB [11], a tool built upon Java, JavaFX, and JavaScript, enables the creation of interactive SVG visualizations for models developed in ProB using an approach that does not require user to have prior knowledge of JavaScript.

Development is ongoing and part of the lead author's postdoc at Brown University, where Sterling's flexible architecture is being leveraged to develop user studies with the goal of better understanding the role of visualization and user interaction in state-based modeling. Additionally, Sterling is the visualizer for an Alloy-like model finder called Forge [1], which is being developed at Brown University and is used to teach a Logic for Systems class of over 60 students.

Acknowledgments. We thank Shriram Krishnamurthi, Tim Nelson, and Kathi Fisler for their ideas and support, Mia Santomauro for the Sterling custom visualization guide, and the Alloy community for their helpful suggestions. This work is partially supported by the US NSF.

References

1. Forge. https://github.com/tnelson/Forge. Accessed 12 Apr 2021
2. Baugh, J., Altuntas, A.: Formal methods and finite element analysis of hurricane storm surge: a case study in software verification. Sci. Comput. Program. **158**, 100–121 (2018)
3. Couto, R., et al.: Improving the visualization of Alloy instances. Electron. Proc. Theor. Comput. Sci. **284**, 37–52 (2018)
4. Dyer, T., Altuntas, A., Baugh, J.: Bounded verification of sparse matrix computations. In: Proceedings of the Third International Workshop on Software Correctness for HPC Applications, Correctness 2019, pp. 36–43. IEEE/ACM (2019)
5. Gammaitoni, L., Kelsen, P.: Domain-specific visualization of Alloy instances. In: Ait, A.Y., Schewe, K.D. (eds.) Abstract State Machines, Alloy, B, TLA, VDM, and Z, pp. 324–327. Springer, Heidelberg (2014). https://doi.org/10.1007/978-3-662-43652-3_33
6. Jackson, D.: Software Abstractions: Logic, Language, and Analysis. The MIT Press, Cambridge (2012)
7. Ladenberger, L., Leuschel, M.: BMotionWeb: a tool for rapid creation of formal prototypes. In: De Nicola, R., Kühn, E. (eds.) SEFM 2016. LNCS, vol. 9763, pp. 403–417. Springer, Cham (2016). https://doi.org/10.1007/978-3-319-41591-8_27
8. Macedo, N., et al.: Sharing and learning Alloy on the web. arXiv abs/1907.02275 (2019)
9. Misue, K., Eades, P., Lai, W., Sugiyama, K.: Layout adjustment and the mental map. J. Vis. Lang. Comput. **6**(2), 183–210 (1995)
10. Oladimeji, P., Masci, P., Curzon, P., Thimbleby, H.: PVSio-web: a tool for rapid prototyping device user interfaces in PVS. Electron. Commun. EASST **69** (2014)
11. Werth, M., Leuschel, M.: VisB: a lightweight tool to visualize formal models with SVG graphics. In: Raschke, A., Méry, D., Houdek, F. (eds.) ABZ 2020. LNCS, vol. 12071, pp. 260–265. Springer, Cham (2020). https://doi.org/10.1007/978-3-030-48077-6_21
12. Zaman, A., et al.: Improved visualization of relational logic models. Technical report. CS-2013-04, University of Waterloo (2013)

Extending ASMETA with Time Features

Andrea Bombarda[1]([✉]) [ID], Silvia Bonfanti[1] [ID], Angelo Gargantini[1] [ID],
and Elvinia Riccobene[2] [ID]

[1] Dipartimento di Ingegneria Gestionale, dell'Informazione e della Produzione,
Università degli Studi di Bergamo, Bergamo, Italy
{andrea.bombarda,silvia.bonfanti,angelo.gargantini}@unibg.it
[2] Dipartimento di Informatica, Università degli Studi di Milano, Milan, Italy
elvinia.riccobene@unimi.it

Abstract. ASMs and the ASMETA framework can be used to model
and analyze a variety of systems, and many of them rely on time constraints. In this paper, we present the ASMETA extension to deal with
model time features.

1 Introduction

Abstract State Machines (ASMs) [8] have been used to model several real case
studies [2,3,5]. The framework ASMETA [4] supports the design and analysis
of ASM models; it offers a wide set of features for model validation, verification, and code generation. However, many real systems, especially those in the
safety-critical and cyber-physic domains, rely on time constraints. Modeling and
validating these kinds of systems using ASMETA may be difficult since it does
not offer primitives explicitly designed for dealing with time. According to the
ASM definition of monitored locations, time is a monitored function whose value
is written by the environment and read by the machine. Till now, the user has
been asked to act as the environment and directly set time values when required;
alternately, boolean monitored functions have been used to model passed time
events. This user-based way of time supplying can be an annoying and error-
prone activity and would require suitable constraints to guarantee time correctness, such as that time is a monotonic increasing function. Moreover, if the
specification uses multiple time units (like seconds and minutes), it is left to the
user to set them in a coherent way.

In this paper, we present the ASMETA library TimeLibrary that introduces *time* as special monitored functions and the concept of *timers*. Moreover,
ASMETA is now extended to handle time in different ways (behind the above-
mentioned already existing ways): *i)* reading the time from the machine hosting
the simulation; *ii)* allowing the user to set the simulation time unit and enter the
time values as a normal monitored function, in case exact time instants chosen
by the user are needed to simulate critical behavior; *iii)* automatically increasing
the time values at each machine step according to parameters initially set by the
user.

© Springer Nature Switzerland AG 2021
A. Raschke and D. Méry (Eds.): ABZ 2021, LNCS 12709, pp. 105–111, 2021.
https://doi.org/10.1007/978-3-030-77543-8_8

```
module TimeLibrary                              definitions:
import StandardLibrary
export *                                        function currentTime($t in Timer) = if (timerUnit($t)=NANOSEC) then
signature:                                          mCurrTimeNanosecs
    abstract domain Timer                           else if (timerUnit($t)=MILLISEC) then mCurrTimeMillisecs
    enum domain TimerUnit={NANOSEC,                 else if (timerUnit($t)=SEC) then mCurrTimeSecs
        MILLISEC, SEC, MIN, HOUR}                   else if (timerUnit($t)=MIN) then mCurrTimeMins
    monitored mCurrTimeNanosecs: Integer            else if (timerUnit($t)=HOUR) then mCurrTimeHours
    monitored mCurrTimeMillisecs: Integer           endif endif endif endif endif
    monitored mCurrTimeSecs: Integer            function expired($t in Timer) = (currentTime($t) >= start($t) + duration($t))
    monitored mCurrTimeMins: Integer
    monitored mCurrTimeHours: Integer           macro rule r_reset_timer($t in Timer) = start($t) :=
    controlled start: Timer—> Integer              currentTime($t)
    controlled duration: Timer —> Integer       macro rule r_set_duration($t in Timer, $ms in Integer) =
    controlled timerUnit: Timer —> TimerUnit       duration($t) := $ms
    derived currentTime : Timer—> Integer       macro rule r_set_timer_unit($t in Timer, $unit in TimerUnit) =
    derived expired: Timer —> Boolean              timerUnit($t) := $unit
```

Code 1. ASMETA TimeLibrary

Our approach is inspired by the timing mechanism provided in other formal notations [9] and in other ASM frameworks. For instance, CoreASM[1] introduces the TimerPattern: it uses the monitored location *now* to save the current system time and has appropriate *TimerAssumptions* on *now* evolution and whatever unit assumptions [8]. However, CoreASM explicitly only links *now* to the machine clock and it manages only times expressed in milliseconds and nanoseconds. Other ASM time mechanisms have been proposed starting from the seminal work in [11]; e.g., a simulator for real-time reactive ASMs was presented in [1], while the TASM approach specifying duration of rule execution appeared in [12], and its extension with events and observers in [13]. A general study of timing for ASMs can be found in [10].

The paper is structured as follows. In Sect. 2 we present the main functionalities we have introduced to deal with time, namely the TimeLibrary, with its *monitored functions* and the *Timer*. Section 3 reports the different approaches to simulate the time and shows the results of simulation in different case studies, such as a simple clock and the well-known Sluice Gate Control case study. Future works are outlined in Sect. 4.

2   Time in ASMETA

In ASMETA framework, we have introduced the TimeLibrary[2] which contains the basic constructs necessary to introduce time features in ASMETA specifications: *i)* monitored functions to manage the time in different time units (nanoseconds, milliseconds, seconds, minutes, and hours); *ii)* an abstract domain Timer useful to introduce user-defined timers; *iii)* some functions and rules to operate on timers, like to check if a desired amount of time is passed, to reset and start a timer, and to set the timer duration and time unit. The proposed solution allows users to use different time units in the same ASM model and it guarantees consistency

---

[1] https://github.com/CoreASM/coreasm.core/tree/master/org.coreasm.engine/src/org/coreasm/engine/plugins/time.

[2] https://github.com/asmeta/asmeta/blob/master/asm_examples/STDL/TimeLibra ry.asm.

between them during model simulation. Moreover, our mechanism assures that in a defined state, all the time functions refer to the same time instant, no matter what time unit is used.

A simple example using the time monitored functions is shown in Code 2, representing a clock that displays at each step current hours, minutes, and seconds.

```
asm simpleClock
import TimeLibrary

signature:
   controlled clockHours: Integer
   controlled clockMins: Integer
   controlled clockSecs: Integer
```

```
definitions:
   main rule r_main =
      par
         clockHours:=mCurrTimeHours mod 24
         clockMins:=mCurrTimeMins mod 60
         clockSecs:=mCurrTimeSecs mod 60
      endpar
```

Code 2. Time example: return current time

Measuring the absolute time is useful, but often, systems require that actions are executed if a desired amount of time is passed. For this purpose, timers are available in the TimeLibrary (see Code 1)[3], and, in the following, we will show how to use them. In the Sluice Gate Control case study (a well-known case study proposed in [7]) there are two timers, one to check if 10 min are passed before closing the gate and one if 3 h are passed before opening the gate. We have declared one function for each timer (see Code 3). Timers are initialized in the initial state, in terms of duration and time unit (using duration($t in Timer) and timerUnit($t in Timer) controlled functions). Moreover, during the initialization phase, the user can (if it is required by the specification) start the declared timer (using the currentTime($t in Timer) library function). The user uses the function expired($t in Timer) from the TimeLibrary (see line 12 in Code 3) to check if the timer passed as parameter is expired. While, when the timer must be reset in the specification, it can be done using the rule r_reset_timer($t in Timer) (see line 15 in Code 3), which takes the timer to reset as parameter. Moreover, in the specification duration of a timer can be changed (using the rule r_set_duration($t in Timer)) as well as its time unit (using the rule r_set_timer_unit($t in Timer)).

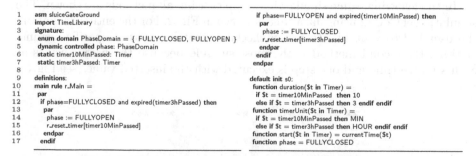

```
1    asm sluiceGateGround
2    import TimeLibrary
3    signature:
4       enum domain PhaseDomain = { FULLYCLOSED, FULLYOPEN }
5       dynamic controlled phase: PhaseDomain
6       static timer10MinPassed: Timer
7       static timer3hPassed: Timer
8
9    definitions:
10   main rule r_Main =
11      par
12         if phase=FULLYCLOSED and expired(timer3hPassed) then
13            par
14               phase := FULLYOPEN
15               r_reset_timer[timer10MinPassed]
16            endpar
17         endif
```

```
         if phase=FULLYOPEN and expired(timer10MinPassed) then
            par
               phase := FULLYCLOSED
               r_reset_timer[timer3hPassed]
            endpar
         endif
      endpar

default init s0:
   function duration($t in Timer) =
      if $t = timer10MinPassed  then 10
      else if $t = timer3hPassed then 3 endif endif
   function timerUnit($t in Timer) =
      if $t = timer10MinPassed then MIN
      else if $t = timer3hPassed then HOUR endif endif
   function start($t in Timer) = currentTime($t)
   function phase = FULLYCLOSED
```

Code 3. Use of Timer in Sluice Gate Control specification

[3] Note that $t denotes the variable t in the AsmetaL notation.

3   Simulating Time

Besides time modeling, ASMETA framework supports three different mechanisms to handle time during simulation: *i)* the time is read from the machine hosting the simulation; *ii)* the user enters the values for time as normal monitored functions; *iii)* the time is automatically increased at each step by a predefined value.

The first mechanism allows the user to run the specification without entering the value of time monitored functions because the time is obtained from the Java 8 Date/Time API *Instant.now()* and automatically assigned to the time monitored functions. Sometimes, and especially if the specification would require long time intervals like hours or very short time intervals like nanoseconds, if the real time is used during the simulation, it may be unfeasible or impractical for the user to check what happens at specific instants of time. In this case, the second mechanism is most suitable: the user specifies the time unit he wants to run the specification and enters the desired time when required. If the specification uses more than one time with different time units, the others are automatically derived. In case the user wants to execute the specification and automatically increment the time by a predefined value at each step, the third approach can be used. The user has to define the time step and time unit, then the system automatically increments the time of the set delta value at each running step. If times have other time units compared to the one set by the user, they are automatically derived. The desired mechanism is set in the ASMETA → Simulator preferences from Window menu in Eclipse, as shown in Fig. 1.

Time mechanism:		
○ use java time ● ask user ○ auto increment		
Delta if auto increment	1	
Preferred time unit	SECONDS	⌄

Fig. 1. Simulator settings

In the following, some simulation examples using all methods are shown. The simulation of Code 2 using Java time is shown in Fig. 2. For the entire simulation, the user had to wait 1 h. To fasten checking what happens at specific instants of time, the second method is the most suitable because the user specifies at each step the time and one step is executed with the inserted value (see Fig. 3).

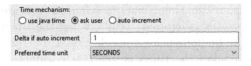

Type	Functions	State 0	State 1	State 2	State 3	State 4	State 5	State 6
C	clockHours		0	0	0	0	0	1
C	clockMins		0	0	1	8	59	0
C	clockSecs		2	25	2	20	18	2

Fig. 2. Clock simulation in "Java Time" mode

Type	Functions	State 0	State 1	State 2	State 3	State 4	State 5
C	clockHours		0	0	0	0	1
C	clockMins		0	0	0	50	0
C	clockSecs		0	1	25	0	0
M	mCurrTimeHours	0	0	0	0	0	1
M	mCurrTimeMins	0	0	0	0	50	60
M	mCurrTimeSecs	0	1	25	3000	3600	

Insert value of monitor function X
Insert a integer constant for mCurrTimeSecs
OK   Clean

Fig. 3. Clock simulation in "ask user" mode

Type	Functions	State 0	State 1	State 2	State 3		State 119	State 120	State 121	State 122	State 123
C	clockHours		0	0	0		0	1	1	1	1
C	clockMins		0	1	1		59	0	0	1	1
C	clockSecs		30	0	30		30	0	30	0	30
M	mCurrTimeHours	0	0	0	0		1	1	1	1	1
M	mCurrTimeMins	0	1	1	2		60	60	61	61	62
M	mCurrTimeSecs	30	60	90	120	...	3600	3630	3660	3690	3720

Fig. 4. Clock simulation in "auto increment" mode with Delta = 30 and Time Unit = SECONDS

Type	Functions	State 0	State 1	State 2	State 3	State 4	State 5
M	mCurrTimeMins	0	75	183	189	193	
C	duration(timer10MinPassed)	10	10	10	10	10	10
C	phase	FULLYCLOSED	FULLYCLOSED	FULLYCLOSED	FULLYOPEN	FULLYOPEN	FULLYCLOSED
C	start(timer10MinPassed)	0	0	0	183	183	183
C	duration(timer3hPassed)	3	3	3	3	3	3
C	timerUnit(timer10MinPassed)	MIN	MIN	MIN	MIN	MIN	MIN
C	timerUnit(timer3hPassed)	HOUR	HOUR	HOUR	HOUR	HOUR	HOUR
C	start(timer3hPassed)	0	0	0	0	0	3
M	mCurrTimeHours	0	1	3	3	3	

Fig. 5. Sluice Gate simulation in "Java Time" mode

The advantage is that the required simulation time is lower. Moreover, this is useful in case the user wants to check the behavior of the modeled system when the clock returns erroneous values (such as decreasing time between two consecutive steps). The last simulation method automatically increments the time at each step by a given value, and an example is shown in Fig. 4. The simulation is performed using a delta time equals to 30 s. As expected the change of clockHours function occurs at State 120. To show the use of timers, we have simulated the Sluice Gate specification using the three methods available. The first method requires the user to wait three hours before changing from FULLYCLOSED to FULLYOPENED. After ten minutes the gate moves back to FULLYCLOSED state in which it remains again for three hours (see Fig. 5). Note that controlled functions at state i are updated due to monitored values (including time) observed at state $i - 1$. Since the instants of time when state changes occur are well known, we have simulated the specification using the second method (see Fig. 6) where the time is set at each state by the user. In this case, the simulation is faster because we do not have to wait the specification

Type	Functions	State 0	State 1	State 2	State 3	State 4	State 5	State 6	Insert value of monitor function ✕
C	phase	FULLYCLOSED	FULLYCLOSED	FULLYCLOSED	FULLYOPEN	FULLYOPEN	FULLYCLOSED	FULLYCLOSED	Insert a integer constant for mCurrTimeMins:
C	duration(timer10MinPassed)	10	10	10	10	10	10	10	[ ]
C	start(timer10MinPassed)	0	0	0	180	180	180	180	OK Clean
C	timerUnit(timer10MinPassed)	MIN	MIN	MIN	MIN	MIN	MIN	MIN	
C	duration(timer3hPassed)	3	3	3	3	3	3	3	
C	start(timer3hPassed)	0	0	0	0	0	3	3	
C	timerUnit(timer3hPassed)	HOUR	HOUR	HOUR	HOUR	HOUR	HOUR	HOUR	
M	mCurrTimeMins	0	10	180	185	190	195		
M	mCurrTimeHours	0	0	3	3	3	3		

Fig. 6. Sluice Gate simulation in "ask user" mode

time. In Fig. 7 the specification is simulated with the auto increment method, the delta is set to 10 and the time unit to minutes. Using this approach, the simulation is run and at each step the minutes are incremented by 10.

4   Future Work

Those presented in this paper are the first results of our effort toward endowing ASMETA with primitives to model and analyze systems with time constraints. In the future, we plan to provide two major improvements: *i)* extending the ASMETA scenario-based validation with new time features; *ii)* automatically mapping ASMETA time primitives into code time primitive, e.g., extending the automatic mapping of ASMETA models into C++ code for the Arduino platform [6].

Type	Functions	State 0	State 1		State 17	State 18	State 19	State 20
C	phase	FULLYCLOSED	FULLYCLOSED		FULLYCLOSED	FULLYOPEN	FULLYCLOSED	FULLYCLOSED
C	duration(timer10MinPassed)	10	10		10	10	10	10
C	start(timer10MinPassed)	10	10		10	180	180	180
C	timerUnit(timer10MinPassed)	MIN	MIN		MIN	MIN	MIN	MIN
C	duration(timer3hPassed)	3	3		3	3	3	3
C	start(timer3hPassed)	0	0		0	0	3	3
C	timerUnit(timer3hPassed)	HOUR	HOUR		HOUR	HOUR	HOUR	HOUR
M	mCurrTimeMins	10	20		180	190	200	210
M	mCurrTimeHours	0	0	...	3	3	3	3

Fig. 7. Sluice Gate simulation in "auto increment" mode with Delta = 10 and Time Unit = MINUTES

References

1. Slissenko, A., Vasilyev, P.: Simulation of timed abstract state machines with predicate logic model-checking. J. Univ. Comput. Sci. **14**(12), 1984–2006 (2008)
2. Arcaini, P., Bonfanti, S., Gargantini, A., Mashkoor, A., Riccobene, E.: Integrating formal methods into medical software development: the ASM approach. Sci. Comput. Program. **158**, 148–167 (2018)

3. Arcaini, P., Gargantini, A., Riccobene, E.: Rigorous development process of a safety-critical system: from ASM models to Java code. Int. J. Softw. Tools Technol. Transf. **19**(2), 247–269 (2015). https://doi.org/10.1007/s10009-015-0394-x
4. Arcaini, P., Gargantini, A., Riccobene, E., Scandurra, P.: A model-driven process for engineering a toolset for a formal method. Softw.: Pract. Exp. **41**, 155–166 (2011)
5. Bombarda, A., Bonfanti, S., Gargantini, A.: Developing medical devices from abstract state machines to embedded systems: a smart pill box case study. In: Mazzara, M., Bruel, J.-M., Meyer, B., Petrenko, A. (eds.) TOOLS 2019. LNCS, vol. 11771, pp. 89–103. Springer, Cham (2019). https://doi.org/10.1007/978-3-030-29852-4_7
6. Bonfanti, S., Gargantini, A., Mashkoor, A.: Design and validation of a C++ code generator from abstract state machines specifications. J. Softw.: Evol. Process **32**(2), e2205 (2019)
7. Börger, E.: Abstract State Machines: A Method for High-Level System Design and Analysis. Springer, Heidelberg (2003). https://doi.org/10.1007/978-3-642-18216-7
8. Böger, E., Raschke, A.: Modeling Companion for Software Practitioners. Springer, Heidelberg (2018). https://doi.org/10.1007/978-3-662-56641-1
9. Furia, C.A., Mandrioli, D., Morzenti, A., Rossi, M.: Modeling Time in Computing. Springer, Berlin Heidelberg (2012). https://doi.org/10.1007/978-3-642-32332-4
10. Graf, S., Prinz, A.: Time in state machines. Fundam. Informaticae **77**(1–2), 143–174 (2007)
11. Gurevich, Y., Huggins, J.K.: The railroad crossing problem: an experiment with instantaneous actions and immediate reactions. In: Kleine Büning, H. (ed.) CSL 1995. LNCS, vol. 1092, pp. 266–290. Springer, Heidelberg (1996). https://doi.org/10.1007/3-540-61377-3_43
12. Lundqvist, K., Ouimet, M.: The timed abstract state machine language: abstract state machines for real-time system engineering. J. Univ. Comput. Sci. **14**(12), 2007–2033 (2008)
13. Zhou, J., Lu, Y., Lundqvist, K.: A TASM-based requirements validation approach for safety-critical embedded systems. In: George, L., Vardanega, T. (eds.) Ada-Europe 2014. LNCS, vol. 8454, pp. 43–57. Springer, Cham (2014). https://doi.org/10.1007/978-3-319-08311-7_5

About the Concolic Execution
and Symbolic ASM Function Promotion
in CASM

Philipp Paulweber[1]([✉]), Jakob Moosbrugger[3], and Uwe Zdun[2]

[1] Vienna University of Technology, Institute of Information Systems Engineering,
Research Unit Compilers and Languages (CompLang),
Argentinierstraße 8, 1040 Vienna, Austria
ppaulweber@complang.tuwien.ac.at
[2] University of Vienna, Faculty of Computer Science, Research Group Software
Architecture (SWA), Währingerstraße 29, 1090 Vienna, Austria
uwe.zdun@univie.ac.at
[3] Vienna, Austria

Abstract. Abstract State Machines (ASMs) are a well-known state
based formal method to describe systems at a very high level and can
be executed either through a concrete or symbolic interpretation. By
symbolically executing an ASM specification, certain properties can be
checked by transforming the described ASM into a suitable input for
model checkers or Automated Theorem Provers (ATPs). Due to the
rather fast increasing state space, model checking and ATP solutions can
lead to inefficient implementations of symbolic execution. More efficient
state space and execution performance can be achieved by using a concolic execution approach. In this paper, we describe an improved concolic
execution implementation for the Corinthian Abstract State Machine
(CASM) language. We outline the transformation of a symbolically executed ASM specification to a single Thousands of Problems for Theorem
Provers (TPTP) format. Furthermore, we introduce a compiler analysis
to promote concrete ASM functions into symbolic ones in order to obtain
symbolic consistency.

Keywords: Abstract State Machine · Concolic execution · CASM ·
TPTP · Z3

1 Introduction

Due to the mathematical foundation of the Abstract State Machine (ASM) theory [1,2], ASM specifications can be evaluated through either concrete or symbolic interpretation. All available ASM implementations offer a concrete execution, and some ASM implementations provide a symbolic execution based on

---

P. Paulweber—The work in this paper was carried out at the former affiliation[2].
J. Moosbrugger—No affiliation.

© Springer Nature Switzerland AG 2021
A. Raschke and D. Méry (Eds.): ABZ 2021, LNCS 12709, pp. 112–117, 2021.
https://doi.org/10.1007/978-3-030-77543-8_9

```
1  CASM
2
3  init test
4
5  [symbolic]
6  function x :  -> Integer
7
8  [symbolic]
9  function y :  -> Integer
10
11  rule test =
12  {
13      if x = 0 then
14          skip
15      else
16          y := 12 / x
17      program( self ) := undef
18  }
19
20
21
22
23  // ...
```

Listing 1.1. Example.casm

```
1  tff(symbolNext , type , sym2: $int).
2  fof(id0,hypothesis,x(1,sym2)).
3  fof('Example.casm:13',hypothesis,sym2=0).
4  fof(id1,hypothesis,x(2,sym2)).
5  fof(final0,hypothesis,x(0,sym2)).
```

Listing 1.2. If-Then-Branch TPTP Trace of Example.casm by Lezuo [6]

```
1   tff(symbolNext , type , sym2: $int).
2   fof(id0,hypothesis,x(1,sym2)).
3   fof('Example.casm:13',hypothesis,sym2!=0).
4   tff(symbolNext , type , sym4: $int).
5   tff(symbolNext , type , sym5: $int).
6   fof(id1,hypothesis,y(1,sym5)).
7   fof(id2,hypothesis,x(2,sym2)).
8   fof(id3,hypothesis,y(2,sym4)).
9   fof(final0,hypothesis,x(0,sym2)).
10  fof(final1,hypothesis,y(0,sym4)).
```

Listing 1.3. Else-Branch TPTP Trace of Example.casm by Lezuo [6]

model checking (e.g. Farahbod et al. [3] for CoreASM). Besides the approaches targeting model checking applications, some ASM implementations transform the specifications into Automated Theorem Provers (ATP) problems to check with off-the-shelve solver tools desired properties (e.g. Arcaini et al. [4] for AsmetaL with SMT solver Yices). A major disadvantage of such techniques is that for rather small ASM specifications, huge ATP input problems are generated which result into large states and long evaluation times of the underlying solver.

To overcome this problem, a *concolic execution* [5] can be used to reduce the number of symbolic path conditions by performing a mixed concrete and symbolic interpretation. Branches inside an evaluation are driven by concrete results and only symbolic states of interest are tracked in the output trace which directly optimizes the results. Therefore, concolic execution [5] trades completeness for computation speed. So far, only Lezuo [6] described a concolic execution approach for ASM specifications. Based on a prototype version of the Corinthian Abstract State Machine (CASM) language[1] [7], the described concolic execution performed a model-to-text transformation by emitting directly multiple Thousands of Problems for Theorem Provers (TPTP) [8] traces of the symbolically executed specification. A downside of Lezuos' [6] approach is that for each conditional rule (path condition) the generated TPTP trace gets forked into an *if-then* and *else* part resulting into two TPTP specifications which are emitted during the symbolic execution of an ASM specification.

Listing 1.1 depicts an example CASM specification consisting of two functions – x and y – and a named rule **test** with a block rule, conditional rule, skip rule, and two update rules. This specification represents the running example which was used by Lezuo [6] to describe a *division-by-zero-free* ASM specification expressed in the latest CASM language syntax. Both functions – x and y – are set explicitly to **symbolic** in order to determine a TPTP trace showing that the function y gets only updated with a non-zero Integer value of function x. Two TPTP traces are generated by using Lezuos' [6] implemented (closed

---

[1] For the CASM syntax description, see: https://casm-lang.org/syntax.

source) symbolic execution. Listing 1.2 depicts the *if-then* part and the Listing 1.3 depicts the *else* part. Based on this traces, a language user can use an external ATP solver Z3 [9] or vanHelsing [10] and prove the *division-by-zero-free* property for the functions y and x by analyzing each TPTP trace.

We present in this paper an improved version of the concolic execution for the (open-source) CASM language and implementation. Based on the concolic execution definition by Lezuo [6], we provide two major improvements in the current presented implementation state: (1) the concolic execution generates a single TPTP trace and does not generate forked TPTP traces for each path condition (see Sect. 2); and (2) a language user only has to set ASM functions of interest to **symbolic** and each ASM function is automatically promoted to **symbolic** if there exists a path which updates that ASM function (see Sect. 3). Furthermore, we do not directly generate TPTP traces through a model-to-text transformation. We have implemented an abstraction of the TPTP model and provided an in-memory model-to-model transformation. This design decision allows us to directly (re)use in the CASM compiler the transformed TPTP instance either for further analysis, in-memory evaluation, or emitting to a textual representation in order to use an external solver.

2   CASM Concolic Execution and TPTP Model

CASM is a concrete ASM implementation with a strongly typed inferred specification language. The concolic execution is implemented as forward symbolic execution by reusing and extending the Abstract Syntax Tree (AST) based concrete execution[2]. Due to the CASM compiler design [11], the symbolic constant, calculation, and environment handling is directly implemented on the CASM Intermediate Representation (IR) level[3]. Our own TPTP implementation[4] supports in-memory model-to-model transformation based on the SMT/SAT solver Z3 [9] to invoke a Z3-based evaluation without external tooling.

Since each ASM function can be explicitly selected to be evaluated as symbolic state (annotation syntax), a complete selection of all available ASM functions inside a specification would enable a full symbolic execution of the provided specification. So far we support all basic ASM rules in the transformation except for symbolic **iterate** rules consisting of symbolic path conditions. Listing 1.4 depicts the same *division-by-zero-free* running example as shown in Listing 1.1 with one small change. In this listing the function y is not explicitly set to **symbolic**, because the function of interest we want to analyze is the function x. Function y gets implicitly set to **symbolic** through a novel compiler analysis pass (see Sect. 3) in order to provide symbolic consistency for the specified update to function y where function x is used in the division operation (see Listing 1.4 at Line 16). Listing 1.5 corresponds to the result TPTP trace of the concolic execution. A first look at this TPTP trace gives the impression that it is longer than

---

[2] For CASM front-end, see: https://github.com/casm-lang/libcasm-fe/pull/206.

[3] For CASM mid-end, see: https://github.com/casm-lang/libcasm-ir/pull/29.

[4] For TPTP model, see: https://github.com/casm-lang/libtptp/pull/5.

```
1  CASM
2
3  init test
4
5  [symbolic]
6  function x :  -> Integer
7
8  // concrete, not set symbolic
9  function y :  -> Integer
10
11 rule test =
12 {
13     if x = 0 then
14         skip
15     else
16         y := 12 / x
17     program( self ) := undef
18 }
```

Listing 1.4. Example.casm

```
1  tff(2,type,'%0':$int).
2  tff(4,type,'%1':$o).
3  tff(6,type,'%2':$int).
4  tff(8,type,'%3':$int).
5  tff(12,type,'%4':$int).
6  tff(15,type,'%5':$int).
7  tff(9,hypothesis,'#div#i':($int*$int*$int)>$o).
8  tff(0,hypothesis,'@x':($int*$int)>$o).
9  tff(1,hypothesis,'@y':($int*$int)>$o).
10 tff(3,hypothesis,'@x'(1,'%0')).
11 tff(5,hypothesis,'%1'<=>('%0'=0)).
12 tff(7,hypothesis,~'%1'=>('@x'(1,'%2'))).
13 tff(10,hypothesis,~'%1'=>('#div#i'(12,'%2','%3'))).
14 tff(11,hypothesis,~'%1'=>('@y'(2,'%3'))).
15 tff(13,hypothesis,'@x'(1,'%4')).
16 tff(14,hypothesis,'@x'(0,'%4')).
17 tff(16,hypothesis,'@y'(2,'%5')).
18 tff(17,hypothesis,'@y'(0,'%5')).
```

Listing 1.5. TPTP Trace of Example.casm

both TPTP traces combined of the previous implementation depicted in Listing 1.2 and Listing 1.3, but besides the path condition fork there is a huge difference in the form of the trace representation itself. Lezuos' [6] implementation uses mixed First Order Form (FOF) and Typed First Order Form (TFF) formulae to represent the state evolving which fully complies to the deprecated TPTP versions before 7.0 [8]. Since the latest major revision 7 of TPTP the mixing of FOF and TFF does not work anymore, because variables and constants in FOF formulae are assumed to be in the same infinite domain, which is not the case for any type in a TFF formulae [8]. The later implies that each variable or constant in a TFF formulae is not equal to any variable or constant in a FOF formula. Therefore, we generate a fully typed TPTP trace by using only TFF formulae in the trace result. A transformed TPTP trace consists of four parts: (1) type declarations for intermediate calculations (see Listing 1.5 Line 1 to 6); (2) language operand definitions (see Listing 1.5 Line 7); (3) all function definitions (see Listing 1.5 Line 8 to 9); and (4) the actual trace itself (see Listing 1.5 Line 10-18). Since in TPTP each variable can only be used once and there exists no notion of time, each ASM function gets mapped to a TPTP predicate with 2 or more arguments where the first argument represents an Integer based time. Similar to the definition by Lezuo [6], we use time at 1 to represent the initialization of ASM functions. Time at 0 equals the termination of an ASM execution. This encoding provides an elegant way to describe start and termination constraints, since the times are known before the concolic execution starts. Furthermore, since CASM supports block rules (parallel execution semantics) and sequential rules (sequential execution semantics) the handling of parallelism is an important issue. The evolving of function states (ASM steps) is encoded in the time value of each function in the first argument. Sequential rule computations which create *pseudo update-sets* [7] are not shown and tracked in the TPTP trace except for the remaining update to functions.

3   ASM Function Promotion and Symbolic Consistency

Due to the possibility that some ASM functions in a CASM specification can be marked as `symbolic`, the concolic execution can reach an interpretation of

```
1  casmi: info: promoting function 'y' to be symbolic, because function is
2  updated with symbolic value.
3  Example.casm:16:8..16:19
4      y := 12 / x
5      ^----------^
```

Listing 1.6. CLI Tool Information of ASM Symbolic Function Promotion

the ASM specification where a symbolic value or calculation could be used in an update rule to a concrete ASM function. This would abort the concolic execution and would lead to an execution error, because the symbolic consistency is violated. Therefore, we implemented a symbolic consistency analysis in the compiler pass pipeline which analyses in advance which concrete ASM functions will be updated by symbolic values. Note that updating a symbolic ASM function with a concrete value (e.g. a numeric value) is possible and does not violate symbolic consistency.

The symbolic consistency pass is an AST-based compiler analysis pass and checks if any function update produces a symbolic conflict. Each function, rule parameters, and expression AST node gets annotated by the analysis which labels the nodes either *symbolic*, *concrete*, or *unknown*.

Depending on the annotated functions through the annotation syntax, all functions are labeled either *symbolic* or *concrete* and all other nodes in the AST are labeled *unknown* at the beginning of the analysis. Since CASM supports named rule calls, each possible rule call hierarchy starting from the init statement has to be evaluated in order to determine symbolic consistency. The analysis derives in a step-by-step manner a Rule Call Graph (RCG) where each callable rule has to go through four states – *init*, *started*, *evaluated*, and *finished*. The resulting RCG is used to derive the final symbolic function promotion which assures symbolic consistency.

We implemented a proper reporting of ASM functions which are promoted to symbolic. Listing 1.6 depicts a console output of our CASM interpreter Command Line Interface (CLI) tool named `casmi`[5] which evaluated in concolic/symbolic execution mode the `Example.casm` specification shown in Listing 1.4 and outputs an information message that function y gets promoted to a symbolic ASM function.

4   Conclusion

In this paper, we describe an improved ASM based concolic execution approach which is implemented for the CASM language and its framework.

Novel about this contribution is that the transformation of an ASM specification towards a TPTP model instance is performed through an in-memory model-to-model transformation which allows either further in-memory analysis, optimization, and evaluation of the TPTP instance or a flexible model-to-text transformation into a TPTP textual representation. Furthermore, the implemented approach only generates a single TPTP trace and promotes non-symbolic

---

[5] For CLI tool `casmi`, see: https://github.com/casm-lang/casmi/pull/12.

ASM functions to symbolic ones if the symbolic consistency is violated which is determined in advance through a symbolic consistency pass.

With our new concolic execution approach we aim at a complete translation validation of the CASM compiler implementation itself by checking each internal transformation step of the intermediate models [11]. Moreover, due to the introduction of state and behavioral separation in the CASM language [12], we are currently investigating the ability of automated semantic checking for imported ASM rules from loaded libraries or modules.

Acknowledgements. We would like to thank Andreas Krall[1] for proof-reading the paper and Emmanuel Pescosta for several concolic execution discussions.

References

1. Gurevich, Y.: Evolving Algebras 1993: Lipari Guide - Specification and Validation Methods, pp. 9–36. Oxford University Press Inc, New York (1995)
2. Borger, E., Raschke, A.: Modeling Companion for Software Practitioners. Springer, Heidelberg (2018). https://doi.org/10.1007/978-3-662-56641-1_9
3. Farahbod, R., Glässer, U., Ma, G.: Model checking CoreASM specifications. In: Proceedings of the 14th International ASM Workshop (ASM 2007). Citeseer (2007)
4. Arcaini, P., Gargantini, A., Riccobene, E.: SMT-based automatic proof of ASM model refinement. In: De Nicola, R., Kühn, E. (eds.) SEFM 2016. LNCS, vol. 9763, pp. 253–269. Springer, Cham (2016). https://doi.org/10.1007/978-3-319-41591-8_17
5. Baldoni, R., Coppa, E., D'elia, D.C., Demetrescu, C., Finocchi, I.: A survey of symbolic execution techniques. ACM Comput. Surv. (CSUR) **51**(3), 50 (2018)
6. Lezuo, R.: Scalable translation validation; tools, techniques and framework. Ph.D. thesis, (2014). Wien, Techn. Univ., Diss
7. Lezuo, R., Paulweber, P., Krall, A.: CASM - optimized compilation of abstract state machines. In: SIGPLAN/SIGBED Conference on Languages, Compilers and Tools for Embedded Systems (LCTES), pp. 13–22. ACM (2014)
8. Sutcliffe, G.: The TPTP problem library and associated infrastructure. J. Automated Reason. **59**(4), 483–502 (2017). https://doi.org/10.1007/s10817-017-9407-7
9. de Moura, L., Bjørner, N.: Z3: an efficient SMT solver. In: Ramakrishnan, C.R., Rehof, J. (eds.) TACAS 2008. LNCS, vol. 4963, pp. 337–340. Springer, Heidelberg (2008). https://doi.org/10.1007/978-3-540-78800-3_24
10. Lezuo, R., Dragan, I., Barany, G., Krall, A.: vanHelsing: a fast proof checker for debuggable compiler verification. In: 2015 17th International Symposium on Symbolic and Numeric Algorithms for Scientific Computing (SYNASC), pp. 167–174. IEEE (2015)
11. Paulweber, P., Pescosta, E., Zdun, U.: CASM-IR: uniform ASM-based intermediate representation for model specification, execution, and transformation. In: Butler, M., Raschke, A., Hoang, T.S., Reichl, K. (eds.) ABZ 2018. LNCS, vol. 10817, pp. 39–54. Springer, Cham (2018). https://doi.org/10.1007/978-3-319-91271-4_4
12. Paulweber, P., Pescosta, E., Zdun, U.: Structuring the state and behavior of ASMs: introducing a trait-based construct for abstract state machine languages. In: Raschke, A., Méry, D., Houdek, F. (eds.) ABZ 2020. LNCS, vol. 12071, pp. 237–243. Springer, Cham (2020). https://doi.org/10.1007/978-3-030-48077-6_17

Towards Refinement of Unbounded Parallelism in ASMs Using Concurrency and Reflection

Fengqing Jiang[1], Neng Xiong[1], Xinyu Lian[1], Senén González[2], and Klaus-Dieter Schewe[1(✉)]

[1] UIUC Institute, Zhejiang University, Haining, China
{fengqing.18,neng.18,xinyul.18,kd.schewe}@intl.zju.edu.cn
[2] TMConnected, Linz, Austria

Abstract. The BSP bridging model can be exploited to support MapReduce processing. This article describes how this can be realised using a work-stealing approach, where an idle processor can autonomously grab a thread from a partially ordered pool of open threads and execute it. It is further outlined that this can be generalised for the refinement of an unboundedly parallel ASM by a concurrent, reflective BSP-ASM, i.e. the individual agents are associated with reflective ASMs, i.e. they can adapt their own program.

Keywords: MapReduce · Work stealing · Reflection · Abstract State Machine · BSP bridging model

1 Introduction

The *bulk synchronous parallel* (BSP) bridging model [6] is a model for parallel computations on a fixed number of processors comprising sequences of alternating computation and communication phases. In a computation phase each processor works independently without any form of interaction until it completes the local computation. When all processors have completed their local computations, they continue with a communication phase to exchange data. With all agents having completed their communication, they return to a new computation phase and thus begin a new superstep. BSP computations are specific concurrent algorithms, and as such they are captured by restricted communicating concurrent Abstract State Machines (ASMs) [2] called BSP-ASMs as shown in [3].

A MapReduce computation comprises a *map* phase processing input data to obtain intermediate key-value pairs, a *shuffle* phase redistributing the data and a *reduce* phase aggregating intermediate key-value pairs to yield the final result. In [3] it was shown how BSP-ASMs can be exploited to specify and analyse MapReduce. Examples how MapReduce is realised based on grounds of the BSP model can be found in [4].

In this short paper we first show that the scheduling effort can be minimised by adopting a work stealing approach as introduced in [1]. This approach places

© Springer Nature Switzerland AG 2021
A. Raschke and D. Méry (Eds.): ABZ 2021, LNCS 12709, pp. 118–123, 2021.
https://doi.org/10.1007/978-3-030-77543-8_10

the decision about the next thread to be executed into the individual agents, i.e. whenever an ASM associated with a processor becomes idle, it can autonomously grab a thread from a partially ordered pool of open threads and execute it. We proceed with an outline of work in progress generalising the approach to a general refinement method for unbounded parallelism in ASMs. The problem is that in an implementation only finitely many processors are available, which suggests to look for a refinement of an ASM by a BSP-ASM. In the general case it becomes necessary that the individual agents are associated with reflective ASMs [5], i.e. they must be able to adapt their own program.

In Sect. 2 we present a specification of MapReduce with work stealing by BSP-ASMs, which is based on the work in [3]. In Sect. 3 we describe our observation that this can be used for a general approach to the refinement of unboundedly parallel ASMs by concurrent, reflective ASMs.

2 BSP-ASMs for MapReduce with Work Stealing

For the processing of MapReduce we can assume a shared environment that is accessible by all processes contributing to the computation. In order to simplify synchronisation among the different processes we assume that there exists an environment host machine H as the center spot of the concurrent computation. For the assignment of tasks we require a task queue TQ holding map and reduce tasks. Map and reduce tasks have a similar structure comprising a job name J, associated (bulk) data D, a function label in $\{Map, Reduce\}$, a function F, a set of indices $I \subseteq \mathbb{N}$, and a processing flag in $\{Undo, Processing\}$.

Map tasks are generated by H as computation tasks and corresponding data is associated with them. For storing intermediate results of map functions we require a reduce container RC. When adding a computation job to H, H will also generate a reduce flag rf and add it to RC. When a reduce flag is completed, H generates a reduce task, adds it to TQ and removes the flag from RC. Reduce tasks are generated only by the reduce container RC, if all map tasks of a job have been completed successfully.

The environment is initialized with data and corresponding user-defined map/reduce functions. The environment-host machine H splits the data into m pieces, then creates and pushes m map tasks into the task queue TQ, and adds a corresponding reduce flag rf into the reduce container RC.

ENVIRONMENT_INITIALIZATION =
    Forall $\langle J \in Job, MapF, ReduceF, Data \rangle$ **Do**
      LET $DataBulks = Split(Data)$
      Forall $DataBulk \in DataBulks$ **Do**
        $Task := \langle J, DataBulk, Map, MapF, i, Undo \rangle$
        $H.TQ := H.TQ \cup \{Task\}$

Further computation tasks are added to the environment before the computation starts by the family of machines.

TASK_ASSIGN =
    If Fetch_Request_from(P_i)

Then $Task := H.TQ.findUndo()$
     $Task.PF := Processing$
If Fail_Information_from(P_i)
Then $Task.PF := Undo$
If Success_Information_from(P_i)
Then $TQ := TQ - \{Task\}$
REDUCE_TASK_GENERATION $=$
     If Receive_from_Map_Task
     Then $rf := RC.find(Job)$
          $rf.DataSet(i) := ReceiveData$
     If Completed(rf)
     Then $Task := \langle Job, DataSet, Reduce, ReduceF, i, Undo \rangle$
          $H.TQ := H.TQ \cup \{Task\}$

Each individual process P_i will run its computation task autonomously. If the program detect its availability P_i will communicate with H and fetch a task T_i from TQ, which is either a map or a reduce task.

$P_i = \ Task := FetchRequest()$
     $Task$ execute
     If Task_Success
     Then Send $SuccessInformation$ to H
          If $Task.FT = Map$
          Then Send result to $H.RC$
               If $Task.FT = Reduce$
               Then OUTPUT
     Else
          Send $FailureInformation$ to H
          Restart/reboot process

In case of a map task the process will execute a map phase. While in this phase, the process applies the given map function to the data sequence resulting in multiple $(Key, Value)$ pairs. When the task is completed successfully, the intermediate result is transferred to H. In case of a reduce task the process will execute a reduce phase. Firstly, the process will sort the $(key, valuelist)$ pairs according to Key. Then it will merge the pairs as the output of the computation phase.

An intermediate result returned to H will be stored into one rf based on the job name. The environment host machine H will monitor, whether each reduce flag is completed after passing back from each map invocation. Since there will be m map operations for a certain assignment in total, only when the reduce flag is loaded with whole m intermediate results, H will generate a reduce task based on the completed rf and add to TQ as well as removing this rf from the reduce container.

3   Reflective Refinement of Unbounded Parallel ASMs

We now look at the use of reflective BSP-ASMs in the refinement of an ASM. Our problem is to deal with (unbounded) parallelism in case of finitely many processors. We identify the agents of the target BSP-ASM with the available processors, so our set of agents will be $\{a_i \mid 1 \leq i \leq k\}$ plus the synchronisation agent a_0 which only executes a SWITCH rule. Whenever a machine encounters a parallel construct, it posts program fragments that need to be executed to a *thread pool*. The thread pool should be partially ordered reflecting that some of the open threads depend on others. Then the gist of the refinement is that rules of ASMs need to be modified such that new threads can be posted to the pool and each agent can fetch his next thread from the pool.

Reflective ASMs as defined in [5] use a 0-ary function symbol *self*, which in each state S takes a tree value representing the current signature and the current rule. In each state S the rule represented by a subtree of $val_S(self)$ is considered not just a tree value, but an executable rule $r(S)$, which is used to yield the next state $S + \Delta_{r(S)}(S)$. For this duality we require functions *raise* and *drop* to switch between these two views: *raise* turns a tree value representing a rule into an executable ASM rule and a tree value representing a signature into a set of locations; *drop* turns rules and locations into tree values [5].

The location *self* can be updated like any other location defined by the current signature, i.e. both the signature and the rule can change in every step. However, to support multiple updates of only parts of the tree, reflective ASMs permit also partial updates. Updates concerning the same location ℓ produced by partial assignments are first collected in a multiset $\ddot{\Delta}_\ell$. If the operators and arguments are compatible with each other, this multiset together with $val_S(f(t_1, \ldots, t_n))$ will then be collapsed into a single update (ℓ, v_0).

In order to turn a BSP-ASM into a reflective BSP-ASM the signature $\Sigma_{i,loc}$ must contain a function symbol $self_i$, and the represented rule must be a BSP rule over Σ_i, in which the process rule $r_{i,proc}$ and the barrier rule $r_{i,comm}$ are represented by subtrees of $val_S(self_i)$. The creation of new threads and the fetching of open threads must then become part of the process rule.

A *thread* ϑ is given by a rule $r(\vartheta)$ together with a set of locations, in which the rule needs to be executed. Let us associate in addition each thread with a unique *identifier* id_ϑ and a set $pred(\vartheta)$ of those identifiers of threads that need to be executed before ϑ. The rule $r(\vartheta)$ can be represented by a tree value $t_{r(\vartheta)}$, and also the required state can be represented by a tree value representing a set of terms. It has to be understood that the terms represented in this tree represent the locations that need to be evaluated to execute the rule. Thus, the terms refer to the locations of the individual ASM associated with a particular agent, so we can add the agent a_i to the tree representation.

Thus, a thread ϑ is represented as a tree value

$$\vartheta = label\_hedge(\texttt{thread}, \texttt{id}\langle id_\vartheta \rangle \, \texttt{rule}\langle r_\vartheta \rangle \, \texttt{agent}\langle a_\vartheta \rangle \, t_{sig} \, t_{pred})$$

with

$$t_{sig} = label\_hedge(\texttt{signature},$$
$$label\_hedge(\texttt{func}, \langle f_1 \rangle \langle a_1 \rangle) \dots label\_hedge(\texttt{func}, \langle f_k \rangle \langle a_k \rangle))$$
$$t_{pred} = label\_hedge(\texttt{pred}, \texttt{id}\langle id_1 \rangle \dots \texttt{id}\langle id_m \rangle)$$

Then the local thread pool $pool_i$ associated with an agent a_i is a set of set of such threads that is also represented by a tree.

It must be possible to make the open threads ϑ with $pred(\vartheta) = \emptyset$ available to other agents as well. This can be done in a communication phase. An agent a_i when executing a step in the computation phase and adding new threads to its thread pool also sets $bar_i := \textbf{true}$ to indicate that it is prepared to enter its communication phase. Likewise, if an agent a_i detects $pool_i = \langle \rangle$, it also sets $bar_i := \textbf{true}$.

When agent a_0 sets $barrier$ to $\textbf{true}$, agents a_i broadcast the identifiers and rules of at most $k - 1$ of their open threads ϑ with $pred(\vartheta) = \langle \rangle$ to the other agents except a_0. On receipt of message containing open threads, an idle agent a_j selects the most appropriate thread adding it to its own thread pool and returning a message to the agent a_i that created the thread. The agent a_i on receiving a selection may acknowledge it by evaluating the terms and sending the data, i.e. the terms plus their values to the agent a_j, and removing the thread from its own pool. In case that a thread has been selected by more than one agent, only one selection is acknowledged, the other one is declined by sending an appropriate message. An agent whose selection has been declined makes an alternative choice until it receives a confirmation.

After receiving a confirmation of a selection or after having sent all acknowledgements an agent sets $bar_i := \textbf{false}$. When agent a_0 sets $barrier$ to $\textbf{false}$, all agents resume their computation phase. So the computation phase of agent a_i must start with an initialisation making a thread from the local thread pool $pool_i$ the new active rule by updating $self_i$ and $pool_i$ (removing the thread), and adding the data associated with it (in case the thread came from a different agent a_j) to the local state.

Let us now consider a single reflective ASM $\mathcal{M}$. So let the signature and rule of $\mathcal{M}$ be represented in a location $self$. We can further turn $\mathcal{M}$ into a BSP-ASM with $\mathcal{M}$ associated with agent a_1, while all other agents a_i ($2 \leq i \leq k$) are associated with an ASM with rule $\textbf{skip}$. We can further assume that the rules r_i take the form $\textbf{choose } y \textbf{ with } \psi(y) \textbf{ do Forall } x \textbf{ with } \varphi(x,y) \textbf{ do } r_i'(x,y)$. So the evaluation of $\psi(y)$ and $\varphi(x, y)$ in the current states yields a set of rules $r_i'(x, y)$ that need to be executed. Assuming that these rules are independent of each other they give rise to new threads to be added to $pool_i$.

Technically, this means that the $\textbf{Forall}$-rule is considered as a rule term, i.e. $drop$ is applied to it. From this term the component rules $r_i'(x, y)$ are extracted and the corresponding threads are added to the thread pool, i.e. the rule is refined by $\textbf{choose } y \textbf{ with } \psi(y) \textbf{ do Forall } x \textbf{ with } \varphi(x, y) \textbf{ do } post(t_{r_i'(x,y)})$, where $t_{r_i'(x,y)}$ is the tree term r_ϑ representing the rule $r_i'(x, y)$ and $post$ is an ASM rule that generates an identifier id_ϑ, computes a tree representation t_{sig} of a bounded exploration witness W from the rule representing the terms that are necessary to evaluate the rule and adds a new thread to $pool_i$.

The first thread in the pool (if non-empty) then becomes the new rule of the agent a_i, which means that the rule is followed by an update of the rule part of $self_i$ together with an elimination of the first element in the local thread pool. The next $k - 1$ threads ϑ_i ($2 \leq i \leq k$) with $pred(\vartheta) = \langle \rangle$ become subject of a posting message using a barrier rule. If there are less than $k - 1$ such threads, all remaining threads are considered.

A thread created by a different agent is usually associated with the transfer of data. It is therefore advisable to first look for new local threads, so we proceed analogously with two decisive differences: (1) As we do not yet know if the conditions $\psi(\boldsymbol{y})$ and $\varphi(\boldsymbol{x}, \boldsymbol{y})$ will be satisfied, we use a conditional rule for the thread with the condition $\psi(\boldsymbol{y}) \wedge \varphi(\boldsymbol{x}, \boldsymbol{y})$; (2) The resulting threads may also depend on the threads that have been created before, so $pred(\vartheta)$ will not be empty. However, this dependence is only partial, and the threads on which the new ϑ depends may have already been executed.

4   Concluding Remarks

In this article we sketched an extension of BSP-ASMs by reflection such that the involved single-agent ASMs can adapt their own programs. We outlined our work in progress how this capability can be used to allow these ASMs to select their next rule from a pool of partially ordered rules. These rules and associated data are produced by the unboundedly many parallel branches of an ASM. In this way we envision to refine ASMs with unbounded parallelism by BSP-ASMs with workstealing modus. The advantage of the approach is that we can dispense with sophisticated scheduling. So far, in this short paper we just illustrated the idea on MapReduce as an example, which greatly benefits from the approach, though it does not require much reflection.

References

1. Blumofe, R.D., Leiserson, C.E.: Scheduling multithreaded computations by work stealing. J. ACM **46**(5), 720–748 (1999). https://doi.org/10.1145/324133.324234
2. Börger, E., Schewe, K.D.: Communication in abstract state machines. J. Univ. Comput. Sci. **23**(2), 129–145 (2017). http://www.jucs.org/jucs_23_2/communication_in_abstract_state
3. Ferrarotti, F., González, S., Schewe, K.D.: BSP abstract state machines capture bulk synchronous parallel computations. Sci. Comput. Program. **184** (2019). https://doi.org/10.1016/j.scico.2019.102319
4. Pace, M.F.: BSP vs. MapReduce. In: Ali, H.H., et al. (eds.) Proceedings of the International Conference on Computational Science (ICCS 2012). Procedia Computer Science, vol. 9, pp. 246–255. Elsevier (2012)
5. Schewe, K.D., Ferrarotti, F.: Behavioural theory of reflective algorithms I: reflective sequential algorithms. CoRR abs/2001.01873 (2020). http://arxiv.org/abs/2001.01873
6. Valiant, L.G.: A bridging model for parallel computation. Commun. ACM **33**(8), 103–111 (1990). https://doi.org/10.1145/79173.79181

The CamilleX Framework for the Rodin Platform

Thai Son Hoang$^{(\boxtimes)}$ ⓘ, Colin Snook ⓘ, Dana Dghaym ⓘ,
Asieh Salehi Fathabadi ⓘ, and Michael Butler ⓘ

ECS, University of Southampton, Southampton, UK
{t.s.hoang,cfs,d.dghaym,a.salehi-fathabadi,m.j.butler}@soton.ac.uk

Abstract. We present the CamilleX framework for the Rodin platform
in this paper. The framework provides a textual representation and persistence for the Event-B modelling constructs. It supports direct extensions to the Event-B syntax such as machine inclusion and record structures, as well as indirect extensions provided by other plugins such as
UML-B diagrams. We discusses CamilleX's design and in particular, its
extension mechanisms and examples of their use.

Keywords: Event-B · Rodin platform · XText · CamilleX

1 Introduction and Motivation

The Event-B modelling method [1] is a discrete state-transition formal modelling
language. The main supporting tool for Event-B is the Rodin Platform (Rodin)
[2], which facilitates the editing of Event-B models and reasoning about them.
Rodin is based on Eclipse and provides an extensible platform via Eclipse's
plug-in mechanism. This is very important for the openness approach to both
Event-B as the modelling method and to the supporting Rodin [11].

Rodin [2] is a supporting platform for Event-B and is developed in Eclipse.
One of the main components of the Rodin Core is the Rodin repository. In
particular, the Rodin repository stores the elements in a tree-shaped structured
database and does not make any assumptions about the elements stored in the
repository. The internal structure of the model repository (called the 'Rodin
repository') was influenced by the choice of using Eclipse as the underlying basis
for Rodin [11]. Essentially, an Event-B model in Rodin is a collection of modelling
elements. The 'syntax' using keywords that users see in the GUI is provided by
the corresponding editors and does not exist in the persisted model. As the
structure is independent of the Event-B modelling language, it makes extending
Event-B straight-forward. Another important component of the Rodin Core is
the Rodin builder which runs the Event-B core tools automatically. While the
design of the Rodin repository aids extensibility, it has other consequences. The
models are persisted in XML files which makes it difficult for humans to read

A. Raschke and D. Méry (Eds.): ABZ 2021, LNCS 12709, pp. 124–129, 2021.
https://doi.org/10.1007/978-3-030-77543-8_11

and understand the model. It is difficult, for example, to compare two versions of a model when using version control tools for collaboration. Moreover, it is challenging to develop a functional modelling user interface, in particular, the editor for the XML files.

Our motivation is to have a true, human readable, text-based persistence of Event-B models which overcomes the limitations of the current modelling user interface.

2 Background

This section provides background information about the CamilleX-relevant technology, namely, the Eclipse Modelling Framework (EMF) and XText.

EMF [10] is an Eclipse-based framework for implementing modelling languages. An abstract syntax is defined by a meta-model and code is then generated to provide a repository for instances of the model. In previous work [9] we have implemented EMF tooling for Event-B models with persistence into the Rodin repository. Many of our plug-ins, including UML-B, are based on this Event-B EMF framework and utilise the extension mechanism that we built into it. The basis of this inheritance structure is the generic meta-class, *EventBElement* which provides facilities for extending the meta-model with new features. The most important of these is the *extensions* containment of *AbstractExtension*. Since this is inherited by all other model element classes, an extension containment can be defined for any kind of concrete model element by subclassing *AbstractExtension* and providing support for persistence, processing and translation as required. The CamilleX tools described herein are based on this EMF meta-model and make use of its extension mechanism, both for syntactic extensions to the modelling language as well as to support model contributions provided by other plugins.

XText [3] is a powerful framework for developing programming languages and domain-specific languages. The input to the framework is a grammar describing the input language and the result of the framework tooling is "a full infrastructure, including parser, linker, typechecker, compiler as well as editing support for Eclipse" [7]. In particular, the editing support generated from XText includes features such as content assist and customisable framework for validation and code generation. Internally XText relies on EMF, e.g., for loading the in-memory representation of any parsed text files. This enables XText models to be used by any other EMF-based tools since the XText grammar can be seen as 'just' an alternative persistence for EMF models.

3 CamilleX

The main aim of the CamilleX framework is to provide text-based serialisation of Event-B models. Furthermore given the existing facilities for Event-B in Rodin, we have the following design principles for CamilleX.

- Reuse the existing Event-B tools of Rodin as much as possible.
- Support direct extension of the Event-B syntax to provide additional features.
- Provide compatibility with other kinds of 'higher-level' models that contribute to the overall model, e.g., UML-B diagrams [8].

Section 3.1 gives an overview of the basic design for the CamilleX framework. We will discuss direct extensions to the Event-B syntax in Sect. 3.2 and indirect extension by plug-ins to contain other kinds of models in Sect. 3.3.

3.1 The Basic Design

CamilleX supports two types of textual files XMachine and XContext, which in turn will be automatically translated to the corresponding Rodin Event-B components (machine and context). The reverse transformation from Event-B to CamilleX is also supported and can be invoked manually as shown in Fig. 1. Note that the representation of CamilleX constructs (XMachines and XContexts), uses an extended Event-B EMF to accommodates Event-B syntax extensions (e.g., machine inclusion and records structure) which are 'flattened' into the (core) Event-B EMF during the automatic translation.

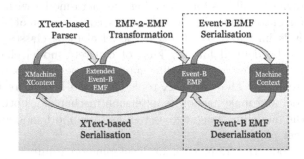

Fig. 1. Overview of CamilleX and Rodin Event-B constructs

Essentially, CamilleX provides the "outer" syntax to Event-B models while relying on the Event-B static checker to check the "inner" syntax of Event-B (i.e., Event-B mathematical formulae).

An important difference between the syntax of CamilleX and that of Camille is that CamilleX supports comments "everywhere". Since Camille relies directly on the structure of the underlying XML serialisation, it can only accept comments attached to the individual modelling elements. For CamilleX, comments can appear anywhere in the textual representation of the Event-B models and are ignored (i.e. omitted) during the translation to Rodin Event-B constructs.

As we rely on the Event-B static checker for checking the inner syntax of the Event-B models, we implemented a callback mechanism to report any errors and warnings raised by the Rodin static checker back to the CamilleX constructs.

3.2   Direct Extensions to the Event-B Syntax

In this section, we present two extensions of the CamilleX constructs to support machine inclusion [6] and records structure [5]. The steps for extending CamilleX are as follows.

1. Extend the Event-B EMF with the new modelling element(s).
2. Extend the grammar of the CamilleX construct and regenerate the supporting tools.
3. Extend the CamilleX validator to ensure the consistency of the added modelling elements.
4. Extend the CamilleX generator to translate the newly added modelling elements into standard Event-B in the model output to Rodin.

Machine Inclusion. The machine inclusion extension provides the concepts for a machine to include other machines and for an event to synchronise with one or more events from the (different) included machines. The details of the mechanism are described in [6]. We summarise the main ideas of for a machine A that includes a machine B below:

- A inherits all variables and invariants of B.
- B's variables can only be modified from A via *synchronising* with an event of B.
  Multiple instances of B can be included via *prefixing* and in that case, B's variables and events are renamed accordingly.

The CamilleX generator is extended to "flatten" the machine inclusion and event synchronisation into standard Event-B EMF before serialisation into Rodin Event-B constructs.

- For each **includes** clause of a machine, the translator copies the variables and invariants from the included machine. If the included machine is prefixed, multiple copies of the variables and invariants are generated and renamed accordingly.
- For each **synchronises** clause of an event, the translator copies the content of the included events (i.e., the parameters, guards and actions) and renames them appropriately if the included machine is prefixed.

Record Structures. The records extension provides the ability to use record structures within Event-B machines and contexts. The record structures and their translation to Event-B are described in [5]. Records can be declared in both contexts and machines and will generate different Event-B modelling elements in each. The CamilleX generator is extended to "flatten" the records into standard Event-B EMF before serialisation into Rodin Event-B constructs. A record in a context will generate either a carrier set or (if it is an inherited record) a constant. The fields of a record in a context will be generated as constants with the appropriate type, depending on the multiplicities, **one** (total functions), **many** (binary relations), or **opt** (partial functions). A record in a machine must inherit another record, and is generates a variable of the machine. The fields of a record in a machine will also be generated as variables with the appropriate type in the machine (depending on the record's multiplicity).

3.3   Indirect Extensions by Plug-Ins

In the previous section, we showed how to directly extend the syntax of the CamilleX constructs, i.e. XContexts and XMachines to support mechanisms such as machine inclusion and record structures. In this section, we describe a generic extensible mechanism for integration with other plug-ins such as UML-B.

We introduce the notion of *containment*, to enable XContexts and XMachine to include external components such as UML-B diagrams. We introduce the contains clauses which references a DiagramOwners. Each DiagramOwner contains zero or more Diagrams which will contribute to the containing Machines or Contexts. The abstract meta-class, Diagram, can then be sub-classed to contribute the specific desired model syntax. For example, the UML-B diagram types, statemachine and classdiagram, both extend Diagram.

An extension point is created for the CamilleX generator, which allows plugins to contribute an implementation of how the contained components are translated in order to contribute to the Event-B models. The CamilleX generator will then defer to this translation for the specific type of contained components declared in the extension, e.g., UML-B state-machines or class diagrams.

4   Conclusion and Future Work

This paper presents the CamilleX framework which provides textual serialisation of Event-B models. In particular, we reuse the existing Event-B tool-chain of Rodin, by providing only the "outer" syntax for the Event-B models. The design of CamilleX supports both direct extension to the Event-B syntax and indirect extensions by plug-ins to contain other types of components such as UML-B diagrams. Our experience shows that CamilleX improves the usability of Rodin and assists users in developing Event-B models.

Future work on machine inclusion will suppress the generation of unnecessary proof obligations (e.g., those that are related to included invariants are correct-by-construction), add support for importing a refinement-chain (instead of individual machines), and integrate with context instantiation.

Currently CamilleX does not fully support the extension refinement of record structures as described in [5]. At the moment, properties of the record fields are translated as axioms and invariants after other "normal" axioms and invariants. Sometimes we need to rearrange the order. For example, the generated elements need to go before or in between the other normal elements. In order to provide more flexible ordering of elements, the Event-B EMF will be restructured to have a single collection of child elements.

Support for reasoning about availability properties with the notion of rigid events and parameters [4] can be also added to CamilleX.

Although the CamilleX containment extension allows for integration with UML-B, the UML-B diagrams are currently persisted in EMF XMI format. For similar reasons to CamilleX, it would be advantageous to have a human-readable text persistence for UML-B diagrams. We are therefore developing XUML-B, which will provide an XText persistence for UML-B.

Acknowledgement. This work is supported by the following projects:

– HiClass project (113213), which is part of the ATI Programme, a joint Government and industry investment to maintain and grow the UK's competitive position in civil aerospace design and manufacture.

– HD-Sec project, which was funded by the Digital Security by Design (DSbD) Programme delivered by UKRI to support the DSbD ecosystem.

References

1. Abrial, J.-R.: Modeling in Event-B: System and Software Engineering. Cambridge University Press, Cambridge (2010)
2. Abrial, J.-R., Butler, M., Hallerstede, S., Hoang, T.S., Mehta, F., Voisin, L.: Rodin: an open toolset for modelling and reasoning in Event-B. Softw. Tools Technol. Transf. **12**(6), 447–466 (2010)
3. Bettini, L.: Implementing Domain-Specific Languages with Xtext and Xtend, 2nd edn. Packt Publishing, Birmingham (2016)
4. Dghaym, D., Hoang, T.S., Butler, M., Hu, R., Aniello, L., Sassone, V.: Verifying system-level security of a smart ballot box. In: ABZ 2021 (2021)
5. Salehi Fathabadi, A., Snook, C., Hoang, T.S., Dghaym, D., Butler, M.: Extensible record structures in Event-B. In: ABZ 2021 (2021)
6. Hoang, T.S., Dghaym, D., Snook, C.F., Butler, M.J.: A composition mechanism for refinement-based methods. In: 22nd International Conference on Engineering of Complex Computer Systems, ICECCS 2017, Fukuoka, Japan, 5–8 November 2017, pp. 100–109. IEEE Computer Society (2017)
7. The XText Project. XText website (2020). https://www.eclipse.org/Xtext/
8. Said, M.Y., Butler, M., Snook, C.: Language and tool support for class and state machine refinement in UML-B. In: Cavalcanti, A., Dams, D.R. (eds.) FM 2009. LNCS, vol. 5850, pp. 579–595. Springer, Heidelberg (2009). https://doi.org/10.1007/978-3-642-05089-3_37
9. Snook, C., Fritz, F., Iliasov, A.: Event-B and rodin documentation Wiki: EMF framework for event-B (2009). http://wiki.event-b.org/index.php/EMF_framework_for_Event-B. Accessed May 2020
10. Steinberg, D., Budinsky, F., Paternostro, M., Merks, E.: Eclipse Modeling Framework. The Eclipse Series, 2nd edn. Addison-Wesley Professional, Boston (2008)
11. Voisin, L., Abrial, J.-R.: The rodin platform has turned ten. In: Aït Ameur, Y., Schewe, K.-D. (eds.) ABZ 2014. LNCS, vol. 8477, pp. 1–8. Springer, Heidelberg (2014). https://doi.org/10.1007/978-3-662-43652-3_1

Extensible Record Structures in Event-B

Asieh Salehi Fathabadi$^{(\boxtimes)}$, Colin Snook , Thai Son Hoang ,
Dana Dghaym , and Michael Butler

ECS, University of Southampton, Southampton, UK
{a.salehi-fathabadi,cfs,t.s.hoang,d.dghaym,m.j.butler}@soton.ac.uk

Abstract. Event-B is a state-based formal method for system development. The Event-B mathematical language does not support a syntax for the direct definition of structured types such as records. This paper proposes extending the Event-B language with direct record definitions. A key feature is the ability to extend records with new fields in refinement steps. The XEvent-B tool, which supports a textual representation of Event-B models, is extended to provide support for direct record definition and automatic transformation of record structures into standard Event-B elements. We demonstrate this work by modelling of the Tokeneer case study.

1 Introduction and Motivation

In Event-B [1], system state is modelled using data structures. However, the Event-B mathematical language does not support a syntax for the *direct* definition of record data structures. Record structures result in more readable models while retaining the ease of refinement and proof. We have extended the Event-B language to support direct definition of record structures. Our motivation is to allow modellers to use the familiar concepts of record structured datatypes in Event-B modelling. Moreover, we aim to have a smooth integration of records with the step-wise refinement paradigm in Event-B. Here, a record is a collection of fields of different data types, in fixed number and sequence [7]. In refinement, we may extend existing records with new fields. This allows us to introduce details to the structured data in an incremental fashion. Our work is inspired by [3] but offers an improved translation into Event-B (see discussion in Sect. 6).

This paper is structured as follows: Sect. 2 provides background on the Event-B. Section 3 presents tool support. Section 4 reports of the syntax and transformation of record structure, followed by application of it in the tokeneer case study presented in Sect. 5. Section 6 compares with other data structuring methods. Finally Sect. 7 concludes.

2 Background

Event-B [1] is a refinement-based formal method for system development. An Event-B model contains two parts: *contexts* for static data and *machines* for

This work is supported by the HiClass project (113213), which is part of the ATI Programme, a joint Government and industry investment to maintain and grow the UK's competitive position in civil aerospace design and manufacture.

A. Raschke and D. Méry (Eds.): ABZ 2021, LNCS 12709, pp. 130–136, 2021.
https://doi.org/10.1007/978-3-030-77543-8_12

dynamic behaviour. Contexts contain carrier sets s, constants c, and axioms $A(c)$ that constrain the carrier sets and constants. Machines contain variables v, invariant predicates $I(v)$ that constrain the variables, and events. Event-B uses a mathematical language that is based on set theory and predicate logic. Event-B is supported by the Rodin Platform [2], an extensible open source toolkit which includes facilities for modelling and verification techniques.

3   Tool: CamilleX

The records feature is based on our EMF framework for Event-B [11] which uses the *Eclipse Modeling Framework* (EMF) [13] and provides extension and translation mechanisms to extend the Event-B language. CamilleX [5] provides an extensible text representation of Event-B models (as opposed to Rodin XML files). CamilleX supports two types of text files, *XMachine* and *XContext*, which are automatically translated to the corresponding Rodin machine or context. We have extended the CamilleX grammar to support the new records extension. CamilleX uses the XText [4] framework to implement an editor and translation tool. XText is an EMF-based open source framework for developing domain-specific languages with a human-readable text persistence. When CamilleX files are saved, the CamilleX translator calls any extension translators. In our case a records translator will generate the 'flattened' standard Event-B elements in the target machine and/or context. Records are translated to standard Event-B and hence direct support for records is not required in the existing Rodin tools.

4   Record Structure

Record Syntax: A record in an Event-B XMachine or XContext text file is specified using the following syntax:

record record_id [extends extended_record_id]
(field_id : [multiplicity] field_type)∗

Each record has an identifier, *record_id*, and can optionally extend another record, *extended_record_id*. A record contains zero or more field(s). A record field has an identifier, *field_id*, an optional multiplicity, *multiplicity*, and a data type, *field_type*.

Multiplicity: Multiplicity defines the minimum and maximum number of times the field element can appear in the record. There are three alternative multiplicity options for a field: - **one**: the field contains exactly one value. - **opt**: (optional) the field contains zero or one values. - **many**: the field contains zero or more values. If no multiplicity option is specified for a field, it is considered as **one**. While the **one** multiplicity is common, our **opt** and **many** multiplicities give modellers the flexibility in defining the cardinality associated with each field in a record.

Extension: A record can be extended via single inheritance, allowing record structures to model hierarchies that occur in refining a model. Instances of an extending record have the fields of the record that they extend as well as the new fields that they define. Static record fields are specified in a context, while dynamic record fields are specified in a machine. A record can extend another record in three ways as follows:

- A record specified in a context/machine extends a record specified in the same context/machine. This approach supports hierarchical definition of data structures. Where some records share some fields, the common fields can be specified as a parent record which is extended by child records.
- A record specified in a machine, extends a record specified in a context, seen by the machine. This is where a record contains both static and dynamic data, we extend fields of a static record by new dynamic fields.
- A record specified in a refining context/machine, extends a record specified in the abstract context/machine. This approach supports data refinement where new fields are defined for an existing abstract record in a refinement level.

Record Transformation: By saving the XText file, a context/machine file including the translated Event-B elements for specified records are generated. The translation elements includes sets, constants and axioms in a context and variables and invariants in a machine. In a context:

- a non-extending record is translated to a set: sets record_id
- an extending record is translated to a constant and an axiom, specifying the record type as a sub-set relation:
  constants record_id
  axioms record_id $\subseteq$ extended_record_id
- each field is translated to a constant and an axiom, specifying the field type:
  constants field_id
  axioms field_id $\in$ record_id ($\leftrightarrow$/$\nrightarrow$/$\rightarrow$) field_type
  There are three alternative relation types for a field depending on its multiplicity: - "many": is translated to a relation ($\leftrightarrow$), - "opt" (optional): is translated a partial function ($\nrightarrow$), - "one": is translated to a total function ($\rightarrow$). In a machine:
- a record in a machine must extend another record. An extending record is translated to a variable and an invariant, specifying the record type as a sub-set relation:
  variables record_id
  invariants record_id $\subseteq$ extendedc_record_id
- each field is translated to a variable and an invariant, specifying the field type:
  variables field_id
  invariants field_id $\in$ record_id ($\leftrightarrow$/$\nrightarrow$/$\rightarrow$) field_type

5   Case Study

The Tokeneer system [8] consists of a secure enclave and a set of system components including a card reader and a fingerprint reader. In this paper, we outline the application of record structures in specifying the system-level states. The primary objective of the tokeneer system is to prevent unauthorised access to the secure enclave. A successful scenario involves: arrival of a permitted user at the door who then presents a card on the card reader and a matching finger print at the fingerprint reader. The system will then unlock the door allowing the user to open it and enter the enclave. A card contains a token and a token includes certificates. Figure 1 presents the hierarchy of certificate types.

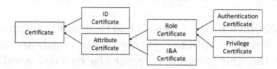

Fig. 1. Hierarchy of tokeneer certificate types

The Event-B model of tokeneer includes an abstract level and two levels of refinement where the door and card specifications are modelled respectively. The certificate hierarchy types, token and card structures are specified using record definitions in the context of the second refinement as below (left); and the machine *m2_card*, seeing context *c2_card*, is partly presented below (right):

```
context c2_card extends c1_door
sets KEYPART PRIVILEGE
     CLEARANCE TOKENID
     FINGERPRINT
records
record CERTIFICATE
   idIssuer: issuer
   validityPeriod: time
   signature: opt KEYPART
record IDCert extends CERTIFICATE
   subject: USER
   publicKey: KEYPART
record AttCert extends CERTIFICATE
   baseCertId: issuer
   tokenId: TOKENID
record RoleCert extends AttCert
   role: PRIVILEGE
   clearance: CLEARANCE
record PrivCert extends RoleCert
record AuthCert extends RoleCert
record IandACert extends AttCert
   fingerprintTemplate: FINGERPRINT
record TOKEN
   tokenID: TOKENID
   idCert: IDCert
   privCert: PrivCert
   iandACert: IandACert
record CARD
   token: TOKEN
record USER extends USER
   fingerprint: FINGERPRINT
end
```

```
machine m2_card refines m1_door sees
     c2_card
variables validToken
records
record USER extends USER
   holds: opt CARD
record TOKEN extends TOKEN
   authCert: opt AuthCert
invariants
@inv1: validToken ⊆ TOKEN
@inv2: holds~ ∈ CARD ⇸ USER
@inv3: ∀tkn. tkn ∈ validToken ⇒
   baseCertId(privCert(tkn)) = idIssuer(
     idCert(tkn)) ∧
   baseCertId(iandACert(tkn)) =
     idIssuer(idCert(tkn)) ∧
   tokenId(privCert(tkn)) = tokenID(tkn
     ) ∧
   tokenId(iandACert(tkn)) = tokenID(
     tkn)
events
event holdCard any user crd where
   @grd1: user ∈ USER
   @grd2: crd ∈ CARD
   @gdr3: user ∉ dom(holds)
   @gdr4: crd ∉ ran(holds)
   @gdr5: token(crd) ∈ validToken
   @gdr6: fingerprint(user) =
     fingerprintTemplate(iandACert(
     token(crd))) then
   @act1: holds(user) := crd end
end
```

Record *USER*, includes the *fingerprint* field, specified in the context, *c2_card*, and is extended in the refining machine, *m2_card*, to include the *CARD* field. We modelled the fingerprint as a static property of a user and holding a card as an optional dynamic property (i.e., defined in the machine). The *holds* field of the *USER* record is declared as optional, specifying that each user can hold at most one card. The invariant *inv2* specifies that each card can be held by at most one user. This is as example of how we can specify extra properties of a defined record.

Record *TOKEN* including a token ID and three static certificates, is extended to include the dynamic optional authorisation certificate in the machine, *m2_card*. During the first attempt to enter the enclave, a valid authorisation certificate is issued and added to the card. A token is valid if all of the certificates on it are well-formed: each certificate correctly cross-references to the ID certificate, and each certificate correctly cross-references to the token ID. This requirement is specified as an invariant, *inv3*; this is an example of the use of records in an invariant.

As an example of modification of the field values, we present the event *holdCard* here. A set of guards check the validity of the card token, *grd5*, and matching fingerprint, *grd6*. If the guards hold then the record field *holds* is updated to include the new pair of *user* and *crd*.

6   Comparison with Other Data Structuring Methods

Alloy. [6] is a lightweight state-based formal modelling language which influenced our approach. Records are similar to Alloy *signatures*, with the same notion of extends and fields. We based our syntax for field multiplicity on Alloy multiplicities (lone=opt, one=one, some=many). Sibling signatures are disjoint by default and signatures can be marked as *abstract* indicating that their *sub-signatures* form a partition. We did not include disjoint/partition features in records because it requires all siblings to be declared simultaneously in the same refinement. *Facts* can be added to constrain signatures with implicit quantification over the instances of the signature. For records, disjointness, partition and constraints can be specified using axioms or invariants.

Declarative Records. A previous attempt [10] to support records (based on [3]) extended the Rodin Event-B notation. Records were converted to 'plain' Event-B by the static checker for proof verification. Records were only available as constants. Hence fields could not be varied individually. In order to change the value of a field, a new instance of the record was selected with the desired new field value and other fields unchanged. However, difficulties were experienced using the ProB model checker since it has to instantiate possible values of records. Our new approach supports variable record structures which alleviates the tooling challenges and our implementation is based on the CamilleX framework to provide human-usable text persistence whereas the previous plug-in persisted models as XML.

UML-B Class diagrams. [9] structure data in a similar way to records. Classes, attributes and associations are linked to Event-B data elements (carrier sets, constants, or variables) and generate constraints on those elements that reflect the data relationships. A class represents a set of instances as does a record and attributes or associations are similar to fields. However, class diagram models provide more options (e.g. multiplicity of fields, partitioning of subsets) than are required in records. Initially, we considered whether a clean human-usable text persistence for class diagrams might provide an efficient route to textual record structures. However, the abstract syntax (meta-model) for class diagrams has been designed to support diagram editors. Since the additional options and structural differences of class diagrams must be accommodated in the persistence, they impact the concrete syntax for records. Therefore, while class diagrams remain an alternative option for data structuring, it is beneficial to provide a separate textual syntax and tooling for records in Event-B.

7   Conclusion and Future Works

In the Event-B formal language, record structures can be defined using standard Event-B elements. For example, a dynamic record field can be specified as a variable and an invariant specifying the type of field. However there is no support for direct definition of record structures. As illustrated by the Tokeneer case study, direct definition of records using our approach results in improved readability in modelling compared with indirect definition in Event-B. In particular, we have designed a new notation for extensible records, so that records can be smoothly extended during the refinement process. However when the Event-B model is refined, records can be refined either by extension (super-position of new fields), presented in this paper, or data refinement (replacement of fields), which is a future research direction. Furthermore, record structures will help identifying program data structures when generating code from Event-B model [12].

References

1. Abrial, J.-R.: Modeling in Event-B: System and Software Engineering. Cambridge University Press, Cambridge (2010)
2. Abrial, J.-R., et al.: Rodin: an open toolset for modelling and reasoning in Event-B. Softw. Tools Technol. Transf. **12**(6), 447–466 (2010). https://doi.org/10.1007/s10009-010-0145-y
3. Evans, N., Butler, M.: A proposal for records in Event-B. In: Misra, J., Nipkow, T., Sekerinski, E. (eds.) FM 2006. LNCS, vol. 4085, pp. 221–235. Springer, Heidelberg (2006). https://doi.org/10.1007/11813040_16
4. Eysholdt, M., Behrens, H.: Xtext: implement your language faster than the quick and dirty way. In: OOPSLA, pp. 307–309. ACM (2010)
5. Hoang, T.S., Snook, C., Dghaym, D., Salehi Fathabadi, A., Butler, M.: The CamilleX framework for the Rodin platform (2021, accepted in ABZ)
6. Jackson, D.: Software Abstractions: Logic, Language, and Analysis. MIT Press, Cambridge (2012)

7. Flatt, M., Felleisen, M., Bruce Findler, R., Krishnamurthi, S.: How To Design Programs: An Introduction to Programming and Computing. MIT Press, Cambridge (2001)
8. Praxis: Tokeneer. https://www.adacore.com/tokeneer. Accessed Mar 2021
9. Snook, C., Butler, M.: UML-B: formal modelling and design aided by UML. ACM Trans. Softw. Eng. Methodol. 15(1), 92–122 (2006)
10. Snook, C.: Event-B and Rodin Wiki: records extension (2010). http://wiki.event-b.org/index.php/Records_Extension. Accessed Mar 2021
11. Snook, C., et al.: Event-B and Rodin Wiki (2009). http://wiki.event-b.org/index.php/EMF_framework_for_Event-B. Accessed Mar 2021
12. Sritharan, S., Hoang, T.S.: Towards generating SPARK from Event-B models. In: Dongol, B., Troubitsyna, E. (eds.) IFM 2020. LNCS, vol. 12546, pp. 103–120. Springer, Cham (2020). https://doi.org/10.1007/978-3-030-63461-2_6
13. Steinberg, D., Budinsky, F., Paternostro, M., Merks, E.: Eclipse Modeling Framework. The Eclipse Series, 2nd edn. Addison-Wesley Professional, Boston (2008)

Formalizing and Analyzing System Requirements of Automatic Train Operation over ETCS Using Event-B

Robert Eschbach$^{(\boxtimes)}$ (ID)

ITK Engineering GmbH, Im Speyerer Tal 6, 76761 Ruelzheim, Germany
robert.eschbach@itk-engineering.de
https://www.itk-engineering.de/en/

Abstract. The European Railway Traffic Management System (ERTMS) aims at the replacement of incompatible national railway traffic management systems in Europe. A part of ERTMS is the European Train Control System (ETCS). ETCS is an automatic train protection system and can collaborate with an automatic train operation system (ATO). ATO can control and monitor the braking, traction and door system of a train. This collaboration is called ATO over ETCS. In this paper we describe the experiences gained in the formalization and the formal analysis of system requirements related to the modes of the ATO onboard unit and its interfaces to train, ATO trackside unit, and ETCS onboard unit. A primary goal to achieve was the stepwise and systematic construction of an Event-B specification tightly coupled with the requirements based on a bidirectional traceability concept. Another goal was the formal verification of important safety properties related to the mode transitions and transition conditions of the ATO onboard unit.

Keywords: ATO over ETCS · Formalization · Traceability · Formal verification · Event-B

1 ATO over ETCS

The European Railway Traffic Management System (ERTMS) aims at the replacement of incompatible national railway traffic management systems in Europe. A central part of ERTMS is ETCS. In this paper we will focus on the collaboration of ETCS with ATO called ATO over ETCS (AoE). The system requirements for ATO over ETCS (AoE) have been specified within the Shift2Rail research project X2RAIL-1. Further information can be obtained from https://projects.shift2rail.org (accessed Feb 11, 2021). The system requirements are specified in SUBSET-125[1]. As can be seen in the reference architecture of

---

[1] Technical specifications for ETCS are published in the Control Command and Signalling Technical Specification for Interoperability hosted by the European Rail Agency. These specifications are grouped into several uniquely numbered subsets.

© Springer Nature Switzerland AG 2021
A. Raschke and D. Méry (Eds.): ABZ 2021, LNCS 12709, pp. 137–142, 2021.
https://doi.org/10.1007/978-3-030-77543-8_13

AoE depicted in Fig. 1 the ATO is divided in ATO onboard unit (ATO-OB) and ATO trackside unit (ATO-TS). The following ATO-OB interface specifications have been analyzed: ATO-TS (SUBSET-125), ETCS onboard unit (ETCS-OB) (SUBSET-130, SUBSET-143) and Train (SUBSET-139). A main part of the system requirements is related to ATO-OB operation modes and mode transitions (see Fig. 2).

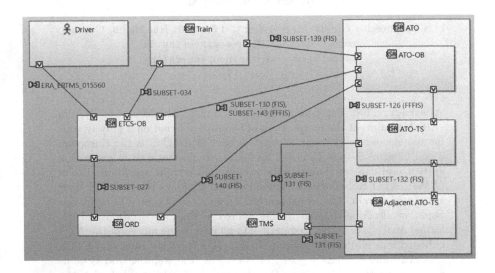

Fig. 1. AoE reference architecture

Related Work. Formal methods have been applied to ETCS in several research and industrial projects. For example, in [9] the authors designed a controller for a cooperation protocol of control parameters. They identified constraints in order to ensure collision freedom. In [7] a concrete implementation of the ETCS Hybrid Level 3 concept is presented. The authors introduce so called virtual block functions which computes the occupation states of virtual subsections. In [3] the authors describe the experiences gained in modelling a satellite-based ERTMS L3 moving block signalling system with Simulink and Uppaal and analysing the Uppaal model with the statistical model checker Uppaal SMC. In [5] the authors present a proof of concept of Virtual coupling. They introduce a specific operating mode for ETCS and define a coupling control algorithm that addresses time-varying delays affecting the communication links. In [1] the authors model the principles of ERTMS Hybrid Level 3 in the mCRL2 process algebra. They perform an analysis with the mCRL2 toolset which can be used for modelling, validation and verification of concurrent systems and protocols. The identified issues have been communicated to the EEIG ERTMS Users Group and have led to several improvements of the affected specifications. In [4] the authors present a detailed report on the convergence of FM related studies carried out in the

Shift2Rail projects X2Rail-2 and ASTRail. In both projects a systematic survey of the state of the art of formal methods application in railway industry and related best practices was carried out. In [2] the authors present the results of a questionnaire which was part of the analysis phase of ASTRail research project. One of the most important results is that classical B and Event-B with the respective tools Atelier-B, ProB and Rodin were used most often in projects.

The formalization approach described in this paper consists of the following six steps (1) identify components, (2) derive function tables for relevant requirements, (3) derive a variable table with read/write permissions for the identified components, (4) derive Event-B specification, (5) validation, and (6) verification.

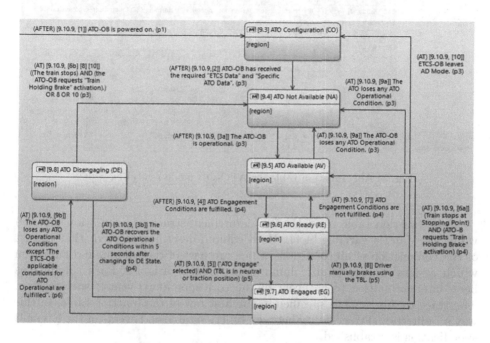

Fig. 2. AoE mode transitions (excerpt)

Step 1: Identify Components. The first step in the formal analysis consists of the systematic identification of components and boundaries specified in SUBSET-125. The ATO-OB is the main component whereas in the environment of ATO-OB the following components were identified: ATO-TS, ETCS-OB, Driver and Train.

Step 2: Derive Function Tables. The formalization started with the stepwise extraction of conditions and actions related to modes like *ATO Engaged* of the ATO-OB and its related mode transitions and transition conditions. For each mode all conditions and actions were combined in a so-called function table as proposed by Dave Parnas (for example [8]). Function tables and Event-B were successfully applied in [6]. Traceability information was added to conditions and actions that allows for relating requirements with elements of the function tables. Furthermore, a coverage analysis was made in order to ensure that all relevant requirements have been covered by the analysis. In Fig. 3 an excerpt of function table for ATO-OB mode ATO Ready (RE) is depicted.

Event-B	[ato_ob_mode_re_3]	[ato_ob_mode_re_4]	[ato_ob_mode_re_5]	[ato_ob_mode_re_6]
	mode = [9.6] ATO Ready (RE)			
	[9.10.8] ATO-OB switched on			
	[9.10.8, 11>p2] The ATO-OB detects no faults which does not allow performing ATO functions (e.g. connection with ETCS-OB is not active while ATO-OB is not in CO State).			
	[9.10.8, 9a>-p3-] The ATO-OB loses any ATO Operational Condition.	[9.10.8, 9a>-p3-] ATO Operational Conditions are fulfilled.		
		[9.10.8, 7>-p4-]ATO Engagement Conditions are not fulfilled.	[9.10.8, 7>-p4-] ATO Engagement Conditions are fulfilled.	
			[9.10.8, 5>-p5-] (Driver has not selected "ATO Engage" OR ETCS-OB is not in AD Mode) OR (TBL is not in neutral AND TBL is not in traction position)	[9.10.8, 5>-p5-] (Driver has selected "ATO Engage" and ETCS-OB is in AD Mode) AND (TBL is in neutral or traction position)
mode	[9.4] ATO Not Available (NA)	[9.5] ATO Available (AV)	[9.6] ATO Ready (RE)	[9.7] ATO Engaged (EG)
train movement	[9.4.1.2] no	[9.4.1.2] no	[9.6.1.4] no	[9.7.1.1] yes

Fig. 3. AoE function table for ATO-OB mode ready (excerpt)

Each function table consists of a hierarchical structured condition block, a value block and a variable block. Each column describes an event that consist of the conjunction of all conditions which are part of the column and an action which assigns in parallel all column values to the variables. Each column has been later formalized to an event of the corresponding Event-B specification. The trace to this event is specified in the first row of the function table. In this way a bidirectional traceability between function table and events of the Event-B specification is established.

Step 3: Derive Variable Table. The next analysis step consists of identifying variables in the conditions and actions so that conditions and actions can be reformulated formally. The requirements were analyzed such that types (like bool or enumeration types) can be assigned to the variables. This involves the analysis of relevant interface specifications, especially the interface between ATO-OB and ETCS-OB and the interface between ATO-OB and train, respectively. Based on this analysis, for each variable and for each component read/write permissions were assigned. In addition, traceability information related to both the interface and system requirements and events of the Event-B specification was added. For example, variable *ato_ob_etcs_ob_in_ad_mode* with type *bool* is defined. The justification for this decision is given by the trace [130, 7.3.2.2] [9.10.8, 10>-p3-] pointing to section 7.3.2.2 of SUBSET-130 and to transition 10 (with priority 3) in section 9.1.8 of SUBSET-125. SUBSET-130 defines a signal with name *Q_ADMODE* which can take the values 0 or 1. The ETCS-OB will send this signal cyclically to the ATO-OB (indicating whether it is in AD mode or not). The component ATO-OB has read permission, the environment component ETCS-OB has write permission. All other environment components have neither read nor write permissions.

Step 4: Derive Event-B Specification. The derived Event-B Specification has 25 variables, 30 invariants and 41 events. In Fig. 4 event `ato_ob_mode_re_6` (excerpt) is shown which has been derived from column [`ato_ob_mode_re_6`] of the function table depicted in Fig. 3.

```
ato_ob_mode_re_6
any
    ato_ob
where
    ato_ob ∈ ATO_OB
    ato_ob_mode(ato_ob) = RE
    ato_ob_operational_conditions(ato_ob) = TRUE
    ato_ob_engagement_conditions(ato_ob) = TRUE
    ato_ob_etcs_ob_in_ad_mode(ato_ob) = TRUE
    ...
then
    ato_ob_mode(ato_ob) = EG
    ...
end
```

Fig. 4. Event ato_ob_mode_re_6

Step 5: Validation. In SUBSET-125 several scenarios of AoE are specified. These scenarios are classified as descriptions (they are not requirements). For example, section 9.10.2 describes the nominal scenario in which the ATO will be engaged by the driver for each so-called journey segment. We have used ProB for simulating step-by-step these scenarios in order to validate the model.

Step 6: Verification. The formal verification was mainly done by theorem proving with the tool Rodin and the Atelier B prover, the SMT solvers and ProB plugins. Furthermore, refinement steps were used to impose further constraints on the environment. For example, since AoE realizes so-called Grade of Automation level 2, a very important safety property is that ATO will never be engaged when the driver never selects ATO Engage. In a refinement step the behavior of the driver was constrained such that he can never select ATO Engage. It was proven that an ATO-OB in this refined Event-B specification is never in mode *ATO Engaged (EG)* or *ATO Disengaging (DE)* (the only modes in which ATO is engaged).

Conclusion. In this paper the experiences gained in the formalisation and analysis of AoE was presented. The formal analysis of the mode transition requirements revealed 23 ambiguities and missing requirements. The formal analysis will be continued in the future. According to the determined refinement strategy the focus will be on the important functional layers ATO Active Functions Table and ETCS mode transition requests (SUBSET-125, 9.11, 9.12), respectively. With these refinement steps the functional layer ATO Operational States (SUBSET-125, 9) will be completely addressed.

Acknowledgments. I thank the anonymous reviewers for their valuable suggestions to improve the paper.

References

1. Bartholomeus, M., Luttik, B., Willemse, T.: Modelling and analysing ERTMS hybrid level 3 with the mCRL2 toolset. In: Howar, F., Barnat, J. (eds.) FMICS 2018. LNCS, vol. 11119, pp. 98–114. Springer, Cham (2018). https://doi.org/10.1007/978-3-030-00244-2_7
2. Basile, D., et al.: On the industrial uptake of formal methods in the railway domain. In: Furia, C.A., Winter, K. (eds.) IFM 2018. LNCS, vol. 11023, pp. 20–29. Springer, Cham (2018). https://doi.org/10.1007/978-3-319-98938-9_2
3. Basile, D., ter Beek, M.H., Ferrari, A., Legay, A.: Modelling and analysing ERTMS L3 moving block railway signalling with simulink and UPPAAL SMC. In: Larsen, K.G., Willemse, T. (eds.) FMICS 2019. LNCS, vol. 11687, pp. 1–21. Springer, Cham (2019). https://doi.org/10.1007/978-3-030-27008-7_1
4. ter Beek, M.H., et al.: Adopting formal methods in an industrial setting: the railways case. In: ter Beek, M.H., McIver, A., Oliveira, J.N. (eds.) FM 2019. LNCS, vol. 11800, pp. 762–772. Springer, Cham (2019). https://doi.org/10.1007/978-3-030-30942-8_46
5. Di Meo, C., Di Vaio, M., Flammini, F., Nardone, R., Santini, S., Vittorini, V.: ERTMS/ETCS virtual coupling: proof of concept and numerical analysis. IEEE Trans. Intell. Transp. Syst. **21**(6), 2545–2556 (2020)
6. Eschbach, R.: Industrial application of Event-B to a wayside train monitoring system: formal conceptual data analysis. In: ter Beek, M.H., McIver, A., Oliveira, J.N. (eds.) FM 2019. LNCS, vol. 11800, pp. 738–745. Springer, Cham (2019). https://doi.org/10.1007/978-3-030-30942-8_43
7. Hansen, D., et al.: Using a formal B model at runtime in a demonstration of the ETCS hybrid level 3 concept with real trains. In: Butler, M., Raschke, A., Hoang, T.S., Reichl, K. (eds.) ABZ 2018. LNCS, vol. 10817, pp. 292–306. Springer, Cham (2018). https://doi.org/10.1007/978-3-319-91271-4_20
8. Parnas, D.L.: Inspection of safety-critical software using program-function tables. In: Linkage and Developing Countries, Information Processing 1994, Proceedings of the IFIP 13th World Computer Congress, Hamburg, Germany, 28 August–2 September 1994, vol. 3, pp. 270–277. IFIP Transactions (1994)
9. Platzer, A., Quesel, J.-D.: European train control system: a case study in formal verification. In: Breitman, K., Cavalcanti, A. (eds.) ICFEM 2009. LNCS, vol. 5885, pp. 246–265. Springer, Heidelberg (2009). https://doi.org/10.1007/978-3-642-10373-5_13

Automatic Transformation of SysML Model to Event-B Model for Railway CCS Application

Shubhangi Salunkhe[1,2]([⊠]) [iD], Randolf Berglehner[1,2], and Abdul Rasheeq[1,2]

[1] DB Netz AG, Frankfurt am Main, Germany
{Shubhangi.Salunkhe-extern,Randolf.Berglehner-extern,
Abdul.Rasheeq-extern}@deutschebahn.com
[2] Neovendi GmbH, Kalkar, Germany
{s.salunkhe,R.Berglehner,a.rasheeq}@neovendi.com

Abstract. Digitalisation and innovation among the railway systems entail effort-demanding challenges, especially when considering how crucial it is to verify safety requirements and proof security levels. The early Verification and Validation (V&V) of railway systems detect critical issues and avoid severe consequences due to software failure. This paper aims to distinguish the subset of SysML language, which can be verified and usable by a systems engineer. As we are interested in proving safety properties expressed using invariants on states, we consider the Event-B method for this purpose. Later the selected SysML subset is used for automatic transformation and finally performing the verification using a formal verification tool. The transformation is applied on a simple point machine case study from DB Netz AG; First, a SysML model is designed using the Windchill modeler, then automatically transformed to Event-B and finally imported into the RODIN platform for formal verification.

Keywords: Model transformation · SysML · Event-B · Model-based systems engineering

1 Introduction

Modern railways are based on centralized control systems where all systems are computer controlled; given the railways' safety-critical nature, modelling and simulating such a system is a high priority. Model-Based Systems Engineering (MBSE) provides generic systems modelling tools that allow users to use this approach in the railway domain. SysML (Systems Modeling Language) [1] is the most widely used general-purpose modelling language in MBSE for specification, analysis, design, and validation of a wide range of different systems. The railway systems are safety-critical; it is essential to implement early Verification and Validation to detect critical issues as early as possible and control the underlying design costs and rollbacks.

© Springer Nature Switzerland AG 2021
A. Raschke and D. Méry (Eds.): ABZ 2021, LNCS 12709, pp. 143–149, 2021.
https://doi.org/10.1007/978-3-030-77543-8_14

This paper proposes a transformation using Triple Graph Grammars (TGGs) [5] between SysML and Event-B [2]. The main idea is to identify the subset of the Event-B language and SysML language, which is necessary and appropriate for the transformation, then search for the semantic similarities between both constructs and finally define a transformation from SysML to Event-B using Model-Driven Engineering techniques. Our work in this paper is an advancement of the MBSE approach explained by Berglehenr *et al.* in [4], and eliminate the manual efforts and increases the efficiency by automating the transformation process.

2   Motivation and Objectives

In the EULYNX[1,2] consortium, European railway infrastructure managers develop standard interfaces and subsystems for the next generation command, control and signaling (CCS) architecture. Model-based systems engineering (MBSE) is used to ensure soundness and completeness of the specified interfaces. Infrastructure managers define the appropriate use case descriptions and modelling experts convert the use cases into executable SysML models using the Windchill Modeler[3] tool. Subsequently, infrastructure managers evaluate whether the specified interfaces are sound regarding their intended use applying simulation-based testing. In order to additionally enable formal verification the SysML model is transformed into a formal Model using the formal method Event-B. The transformation is done manually without tool support.

In the existing formal verification of EULYNX models, the SysML state machines created using Windchill modeler are transformed to equivalent UML-B state machines using UML-B plug-in [7]. Later from the UML-B state machines, the Event-B code is generated, and then the safety requirements are verified. The overall approach is time-consuming and increases the overall life-cycle cost. These challenges motivate our work in this paper to establish the automated transformation with the following set of objectives: (1) The main objective is to propose a methodology and tool-chain to automate the transformation of SysML specification models into formal models (Event-B). (2) The traceability should be maintained between informal requirements and the modeled system, specifically for the safety properties. (3) The model should be verified against such safety requirements using formal methods with some tool support. (4) Reduce the efforts involved in the manual transformation of the SysML semi-formal model to a formal model.

---

[1] https://www.eulynx.eu/.

[2] Our work is featured in embedded video on EULYNX website: https://www.youtube.com/watch?v=GhoNoMm4om0.

[3] https://www.ptc.com/en/products/windchill/modeler.

3   Case Study and Transformation Approach

3.1   Case Study and Scope

In this paper, we have proposed a prototype to automate the transformation. For this prototype, we have considered a small case study provided by DB Netz. It is a technical specification and requirements document describing an interface between a point machine controller to the interlocking. Figure 1 shows Point Machine's configuration in its simplest form: Two tracks represent the left and right positions. The lamps represent the position of the tracks after the movement. The main requirement is to move the tracks to the left or right position depending on the commands received from the interlocking. Figure 2 depicts the system's behavior in the form of a SysML state machine diagram.

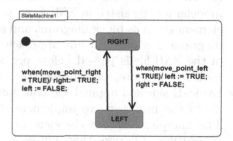

Fig. 1. Case Study: Point Machine   **Fig. 2.**   Case Study: SysML State-Machine Diagram

To perform the formal verification, the SysML state machine diagram is transformed into the equivalent Event-B model. Hence, in the Event-B method, there are mainly two aspects that need to be considered: What kinds of Event-B components? Furthermore, what kind of link between these components to use?. We restrict the Event-B method usage to one machine with variables, invariant, events, actions and guards. Concerning the data type, the primitive data type (i.e. Boolean) is considered. After Studying the Event-B language thoroughly, we define the SysML subset under the defined limited subset for Event-B. Table 1 list the semantic mapping between SysML and Event-B.

We have considered a very limited subset of state machine elements in our approach to provide a prototype as a first step.

3.2   Model-to-Model Transformation

The transformation defined in our approach consists of **TGG** rules [5] that need to be established by using the subset defined for semantics between SysML and Event-B Model Using Model-Driven Engineering (MDE) techniques, we implemented a proprietary model transformation written in eMoflon-IBeX [3].

Table 1. List of semantic similarities

SysML concept	Event-B concept
State-Machine	Machine(Project)
States	Variables
Transition	Events
Effects	Actions
Triggers	Guards
States	Default Invariants
Ports	Variables

As stated earlier, the system functional requirements are modeled in Windchill modeler and persisted in XMI format. The XMI file includes other SysML model elements such as block diagrams and stereotypes, but we focus on state machine diagrams and ports in our approach. Thus, the state machine and ports part of the XMI file is parsed before providing it as an input to the eMoflon: IBeX tool. Furthermore, we consider state machine as a collection of *variables* at the Event-B side and applied the transformation accordingly.

In the first step, we implement the meta-models for SysML and Event-B. The mapping between the elements of SysML and Event-B models is defined by writing TGG rules that will generate the target Event-B model. TGGs can perform the transformation in both direction i.e. *forward* and *backward*, hence named *bi-directional*. In our approach, we only apply *forward* transformation. Listing 1.1 depicts the Event-B model generated by the application of the rules for the state machine depicted in Fig. 2.

Each state machine of SysML model is transformed into Machine module of the Event-B and has the same name as state machine. The States and Ports are transformed into Event-B Variables, for example, states *RIGHT and LEFT* and ports *right, left*. States from SysML state machine are also translated to Invariants (for e.g., TYPEOF_RIGHT:) and initialized them in the *INITIALISATION* event (for e.g., init_RIGHT:RIGHT := TRUE). The predicates of Invariants are created using *attribute conditions* (for e.g., RIGHT∈BOOL). The *attribute conditions* are implemented using *JAVA* language. The Transitions, Triggers and Actions of SysML state smachine are translated into Event-B Events, Guards and Actions. In Event-B, the Event represent the transition from one state to other state depending on the data that represent the state. Such *transitions* are realized by translating the Source State and Target State of a Transition of SysML state machine to Guard condition and Enter & Leave actions of Event-B (for e.g., isin_RIGHT: RIGHT = TRUE & Enter_LEFT: LEFT := TRUE). This prototype has implemented 14 rules and 13 attribute conditions to transform state machine from a case study.

```
MACHINE
  machine
VARIABLES
  right
  move_point_right
  move_point_left
  LEFT
  RIGHT
  left
INVARIANTS
  TYPEOF_LEFT :        LEFT∈BOOL
  TYPEOF_RIGHT :       RIGHT∈BOOL
EVENTS
  INITIALISATION ≙
  STATUS
    ordinary
  BEGIN
    init_RIGHT :     RIGHT := TRUE
    init_LEFT :     LEFT := FALSE
  END
  Previous≙
  STATUS
    ordinary
```

```
  WHEN
    isin_LEFT :       LEFT = TRUE
    guard2 :     move_point_right = TRUE
  THEN
    enter_RIGHT :      RIGHT := TRUE
    action4 :    left := FALSE
    leave_LEFT :     LEFT := FALSE
    action3 :    right := TRUE
  END
  Next≙
  STATUS
    ordinary
  WHEN
    guard1 :     move_point_left = TRUE
    isin_RIGHT :      RIGHT = TRUE
  THEN
    action1 :    left := TRUE
    enter_LEFT :     LEFT := TRUE
    action 2 :   right := FALSE
    leave_RIGHT :     RIGHT := FALSE
  END
END
```

Listing 1.1. Event-B model obtained for the Point Machine State-Machine

In order to perform the transformation, the following steps are performed using *eMoflon:IBeX* tool: (1) First, The SysML state machine model is exported from the Windchill modeler in *.xmi* format and provided as an input to the tool. (2) later the *forward operation* of *eMoflon:IBeX* is applied together with the rules. (3) The transformation will generate the Event-B model in *.xmb* format, which is imported into the RODIN [6] platform to generate the Event-B code.

4   Related Work

Several well-known approaches offer a transformation from SysML to formal methods in different industries according to their needs and applications. Snook and Butler [7] perform a transformation from UML-B to Event-B, where UML-B is a graphical front end for Event-B. Similarly, Bousse *et al.* [8] worked on transformation from SysML to B method, with the same motivation as ours for early V&V of railway systems, but with a different solution. Many other works offered a transformation from SysML to different other formal methods such as Wang *et al.* [9] worked on transformation from SysML to NuSMV model checker. Zhang *et al.* [10] introduced transformation from SysML requirement diagram to Event-B for distributed systems. The seminal work by Giese *et al.* [11] provided a transformation from SysML to AUTOSAR using Triple Graph Grammars (TGGs). Bouwman and Djurre van der Wal *et al.* [12] automated a transformation from SysML to mCRL2 in a project called FormaSig[4] in collaboration with EULYNX, DB Netz and ProRail[5].

Despite the availability of different transformation for SysML to formal methods, the researchers are still trying to perform and implement SysML transformation according to the usage of formal language in their industries, specific to their application, availability of tools, and expertise. The works mentioned in this section are tailored to the specific application scenarios and cannot be

---

[4] https://www.utwente.nl/en/eemcs/fmt/research/projects/formasig/.

[5] https://www.prorail.nl/.

related to a transformation to the Event-B, which motivates our work in this paper to perform the transformation from SysML state machine to Event-B.

5   Conclusion and Future Work

Our approach provides a technique for early V&V of SysML model using the existing tool-set of Event-B language. The essential purpose is to propose a methodology and tool-chain to support the automatic transformation and allow standardization of system interfaces' specifications. The proposed approach is implemented in a prototype using a simple case study. The existing tools such as Windchill modeler for semi-formal modelling (i.e. SysML) and RODIN for formal modelling(i.e. Event-B) are connected via eMoflon: IBeX to perform the transformation. The prototype is limited to the simple and most relevant construct of SysML state machine, in which we performed the semantics mapping between both languages and defined the transformation rules. The prototype is validated by providing three different test cases. The test cases are simple and have the language elements that are described in Sect. 3.1, but they consist of more states, transitions, events and actions as compared to the case study provided in Sect. 3.1. The transformation provides the correct results for all three test cases, thus validating the transformation to be accurate.

The prototype we implemented in this paper acquire elementary constructs of the SysML state machine. We plan to extend the transformation for more complex state machine constructs to provide the practical applicability of the transformation to the real-world models (e.g. EULYNX models). The current transformation is applied in only *forward* direction. The transformation shall be extended for *backward* direction to facilitate traceability in case of error found during formal verification. The RODIN platform and eMoflon: IBeX uses the same EMF framework for their application; in future, these two tools can be merged, and a User Interface(UI) can be implemented to perform the transformation. The UI will help inexperienced users conduct the transformation very easily and liberate them from any technical details.

References

1. Friedenthal, S., Moore, A., Steiner, R.: OMG systems modeling language (OMG SysML) tutorial. In: INCOSE International Symposium, vol. 9 (2006)
2. Abrial, J.-R., et al.: Rodin: an open toolset for modelling and reasoning in Event-B. Int. J. Softw. Tools Technol. Transfer 12(6), 447–466 (2010)
3. Weidmann, N., et al.: Incremental bidirectional model transformation with eMoflon: IBeX. In: Cheney, J., Ko, H.-S. (eds.) 2019 (CEUR Workshop Proceedings), vol. 2355, pp. 45–55. CEUR-WS.org (2019)
4. Berglehner, R., Rasheeq, A., Cherif, I.: An approach to improve SysML railway specification using UML-B and EVENT-B. Poster Presented at RSSRail (2019)
5. Schürr, A.: Specification of graph translators with triple graph grammars. In: Mayr, E.W., Schmidt, G., Tinhofer, G. (eds.) WG 1994. LNCS, vol. 903, pp. 151–163. Springer, Heidelberg (1995). https://doi.org/10.1007/3-540-59071-4_45

6. Butler, M., Hallerstede, S.: The Rodin formal modelling tool. In: FACS 2007 Christmas Workshop: Formal Methods in Industry, pp. 1–5 (2007)
7. Snook, C., Butler, M.: UML-B and Event-B: an integration of languages and tools (2008)
8. Bousse, E., et al.: Aligning SysML with the B method to provide V&V for systems engineering. In: Proceedings of the Workshop on Model-driven Engineering, Verification and Validation (2012)
9. Wang, H., et al.: Integrating model checking with SysML in complex system safety analysis. IEEE Access **7**, 16561–16571 (2019)
10. Zhang, Q., Huang, Z., Xie, J.: Distributed system model using SysML and Event-B. In: Gu, X., Liu, G., Li, B. (eds.) MLICOM 2017. LNICST, vol. 226, pp. 326–336. Springer, Cham (2018). https://doi.org/10.1007/978-3-319-73564-1_32
11. Giese, H., Hildebrandt, S., Neumann, S.: Towards integrating SysML and AUTOSAR modeling via bidirectional model synchronization. In: MBEES (2009)
12. Luttik, S.P.B.: What is the point: formal analysis and test generation for a railway standard. e-proc (ESREL2020 PSAM15) (2020)

Short Articles of the PhD-Symposium
(Work in Progress)

Formal Meta Engineering Event-B: Extension and Reasoning The *EB4EB* Framework

Peter Riviere[✉]

IRIT/INPT-ENSEEIHT, 2 rue Charles Camichel Toulouse, 31000 Toulouse, France
`peter.riviere@toulouse-inp.fr`

1 Context

State-based Formal methods have been used to design and verify the development of complex software systems for a long time. Such methods are underpinned with solid mathematical concepts. Event-B [1] belongs to this family of methods. It advocates a correct-by-construction approach to model a complex system. It is based on set theory and first-order logic. It comes with a powerful integrated development environment called Rodin [9].

The use of formal methods must satisfy the needs of the end user by allowing for scalability, portability, expressiveness, and modularity, among other things. Many key features are currently supported by the Event-B language either in the core modelling language or through specific plugins (e.g. composition plug-in [10], Theory plug-in [2,4], code generation plug-in [6,8]). The Theory Plug-in [2,4] extends Event-B to allow for the definition of new data types, theories, and operators in order to enhance the expressiveness of the formalism. For example, to handle the continuous behavior of a hybrid system for designing a safe controller, domain specific features, related to continuous mathematics [5], have been developed in the form of theories.

The Event-B language requires advanced modelling and reasoning concepts in order to capture the notion of model, proof obligation (PO) and proof process. Currently, verifying interesting properties such as deadlock freeness, event scheduling, liveness, etc., requires ad hoc modelling by the designer. Establishing these properties is based on the use of automatic and/or interactive proof systems and/or model checkers.

Due to a lack of access and explicit manipulation of Event-B concepts, it is quite impossible to express a generic property on these concepts in a theory within a generic definition. Indeed, there is no mechanism in Event-B allowing a designer to define, at a higher order level, additional reusable POs.

The above mentioned issue has been addressed by the development of several plug-ins as Rodin tools. Examples are machine compositions and decomposition, event scheduling, code generation, and translation that have been developed using Eclipse. There is a lack of evidence to guarantee the functional correctness of such developed tools. For example, how can we assert that the machine composition and decomposition plug-in behave correctly.

A. Raschke and D. Méry (Eds.): ABZ 2021, LNCS 12709, pp. 153–157, 2021.
https://doi.org/10.1007/978-3-030-77543-8_15

2   Motivation and Objectives

The above mentioned plug-ins enrich Event-B by introducing either new data types (e.g. using the theory plug-in [2,4]) or by externally defined specific programs that manipulate Event-B models (e.g. composition/decomposition plug-in). None of the above approaches allow to manipulate Event-B models as first order objects. A typical example is the case of deadlock freeness. Three possible options are currently possible: either the developer writes explicitly this PO in the form of a theorem to be proved in the development, or by writing an external program in the PO Generator (to generate this theorem as a PO), or in the form of a plug-in (generating an Event-B machine with the PO as theorem). Both approaches are error prone (written programs are not certified to be correct) or ad hoc (no reuse, properties need to be written for each specific model analysis).

Offering the capability to manipulate Event-B concepts (models, states, transitions, invariants, variants, guards, POs, etc.) as first order objects will allow the developer to express properties on these objects. For example, deadlock freeness PO can be expressed at machine level if guards and invariants can be manipulated in the modelling language. Such a manipulation is possible if the Event-B theory, as defined in the Event-B book [1] associated to Event-B, is formalised in Event-B itself (reflexive modeling).

So, the objective of our PhD thesis work consists in improving modelling and reasoning capabilities of Event-B through the development and formalisation of a theory of Event-B in Event-B. Indeed, we propose to develop the *EB4EB* framework grounded on a set of theories defining data types for Event-B concepts, operators manipulating these concepts and a set of proved theorems precising their semantic properties. In addition, we propose to build other theories to introduce other Event-B models domain specific analyses in the form of properties expressed on the Event-B concepts that describe POs on the analysed models.

Note that the soundness of this framework for Event-B extension and reasoning developed in Event-B shall preserve the core logical foundation of original Event-B models. The main objectives of our work are summarised as follows:

- Analyse and identify the fundamental Event-B concepts and properties that define the notion of machine. Then, using a context or a theory, formalise these concepts (modelling in the small).
- Analyse and identify the Event-B refinement operation for events, data and machine, and then formalise it as a context or theory (modelling in the large).
- Deploy the proposed approach for enhancing reasoning mechanism like deadlock freeness, reachability, etc.
- Introduce new modeling and reasoning mechanisms to handle domain specific analyses of Event-B models. For example, continuous behaviour, human machine interaction, and so on.
- Deploy the proposed approach for analysing and certifying the existing plug-ins, such as composition/decomposition, code generation, etc.

– Allow the capability to Import/Export Event-B models as First Order Logic
  formulas in other proof tools, as these models become expressed as instances
  of Event-B theories.

3 Proposed Approach

3.1 Overview of the Approach

An *Event-B system model* consists of *context, machine* that focuses on formal
modelling to describe system behaviour using refinement approach. Additional
theories may be required to axiomatise new definitions and data-types in either
contexts or *theory* components.

Our approach focuses on the development of Event-B theory axiomatising
Event-B concepts. We propose a set of datatypes, operators, and theorems to
specify the Event-B concepts, their relationships and other additional attributes
and properties related to these concepts. The obtained meta-theory serves to
design a system model as instances of this meta-theory. This instantiation generates a set of new POs.

3.2 Modelling and Instantiation Mechanism

Once the theory for Event-B concepts is designed, two main approaches to instantiate it are envisioned, namely deep modelling and shallow modelling.

– **Deep modelling.** All the concepts of an Event-B model to be analysed, variables, events, guards, invariants, substitutions, etc., are defined as instances
  of the developed meta-theory. An Event-B model is represented as an Event-
  B context, and POs are described as either theorems or well-definedness POs.
  Higher order logic and set theory are used to express all the definitions. It can
  be used as an entry point to the design of an import/export system between
  other proof assistants.
– **Shallow modelling.** In this case, for each Event-B model to be analysed, we
  define another Event-B model consisting of a context instantiating the theory
  with the concepts of the analysed model and a refinement of an abstract
  generic machine composed of two events *init* and *progress* capturing generic
  behaviours. This machine is refined to introduce the events of the analysed
  Event-B model. This method contributes to reduce the proof effort as the
  POs associated each event become simulation proofs. In the same spirit as
  TLA$^+$ [7], the concepts of *init* and *progress* events are identified similar to
  init and *next* events of TLA$^+$.

These two instantiation mechanisms extend the modeling and reasoning capabilities of Event-B language itself as it makes it possible to define additional
theorems encoding other POs (e.g. deadlock freeness). It is important to note
that these two instantiation mechanisms are distinct and play an important role
in the refinement process. The modeling tool Rodin equipped with proving tools

will be used to support all the theories and models. To check the correctness of the developed models, all the generated POs related to well-defined conditions, theorems and properties must be successfully discharged. Instantiation also generates some new POs that must be discharged before any further development. In addition, the developed theory of Event-B can be used for analysing and verifying the core functionalities of existing Rodin plugins.

4   Future Work

The development of Event-B theory is currently in progress. In addition to the developments of the necessary theories, we intend to develop complex case studies to demonstrate the expressiveness and scalability of both deep and shallow mechanisms. Other planned work includes checking the correctness of existing plug-ins like composition/decomposition, code generation, etc., by describing, at the level of the Event-B theory, the operation they encode. Moreover, we plan to deploy the proposed approach for enhancing the reasoning mechanism, such as deadlock freeness, reachability etc. In addition, the proposed approach will be implemented with the theory plug-in and context instantiation developed in the context of the EBRP project. Our long term future work includes to import/export the Event-B theory as well as Event-B models in other proof assistants, through Dedukti [3].

Acknowledgements. This study was undertaken as part of the EBRP (Enhancing EventB and RODIN: EventB-RODIN-Plus) project. We are very grateful to EBRP project members for their valuable discussion and feedback.

References

1. Abrial, J.R.: Modeling in Event-B: System and Software Engineering. Cambridge University Press, Cambridge (2010)
2. Abrial, J.R., Butler, M., Hallerstede, S., Leuschel, M., Schmalz, M., Voisin, L.: Proposals for mathematical extensions for Event-B. Technical Report (2009)
3. Boespflug, M., Carbonneaux, Q., Hermant, O., Saillard, R.: Dedukti: a Universal Proof Checker. In: Journées communes LTP - LAC. Orléans, France, October 2012. https://hal-mines-paristech.archives-ouvertes.fr/hal-01537578
4. Butler, M., Maamria, I.: Mathematical extension in Event-B through the rodin theory component (2010)
5. Dupont, G., Aït-Ameur, Y., Pantel, M., Singh, N.K.: Formally verified architecture patterns of hybrid systems using proof and refinement with Event-B. In: Raschke, A., Méry, D., Houdek, F. (eds.) ABZ 2020. LNCS, vol. 12071, pp. 169–185. Springer, Cham (2020). https://doi.org/10.1007/978-3-030-48077-6_12
6. Fürst, A., Hoang, T.S., Basin, D., Desai, K., Sato, N., Miyazaki, K.: Code generation for event-B. In: Albert, E., Sekerinski, E. (eds.) IFM 2014. LNCS, vol. 8739, pp. 323–338. Springer, Cham (2014). https://doi.org/10.1007/978-3-319-10181-1_20
7. Lamport, L.: Specifying Systems, The TLA+ Language and Tools for Hardware and Software Engineers. Addison-Wesley Longman Publishing Co., Inc. (2002)

8. Méry, D., Singh, N.K.: Automatic code generation from Event-B models. In: Proceedings of the 2011 Symposium on Information and Communication Technology, SoICT 2011, pp. 179–188 (2011)

9. Rodin sourceforge. https://sourceforge.net/projects/rodin-b-sharp/

10. Silva, R., Butler, M.: Shared event composition/decomposition in Event-B. In: Aichernig, B.K., de Boer, F.S., Bonsangue, M.M. (eds.) FMCO 2010. LNCS, vol. 6957, pp. 122–141. Springer, Heidelberg (2011). https://doi.org/10.1007/978-3-642-25271-6_7

A Modeling and Verification Framework for Security Protocols

Mario Lilli[(✉)] [ID]

Computer Science Department, Università degli Studi di Milano,
via Celoria 18, Milan, Italy
mario.lilli@unimi.it

Abstract. Bad design decisions in security protocols can drastically affect the robustness of the protection given by protocols, causing the introduction of vulnerabilities and leak of information. My PhD project aims to reduce the possibility of introducing flaws supporting designers and engineers with a user-friendly formal verification framework, with various options for both model construction and verification.

Keywords: ASM · Security protocol · Automatic verification · ASMETA

1 Introduction

A challenging problem in security protocols design is the difficulty of granting security properties in a provable way. This problem can be traced back to both a lack of user-friendly tools and protocol designers' misconceptions linked to the formal methods [7]. During my PhD that has started in October 2020, I aim to study and develop a new tool that hides mathematical formalism and enforces push-button verification. Many other tools address the same problem [2,3,6,9], but no one integrates all the required features to lead the protocol designers to consider the formal methods a viable way to check their protocol's correctness. The common problems of these tools are: a difficult modelling language, making the writing of the model error-prone as well; a verification process that might require user interaction and knowledge of the tool's internal; the verification results difficult to interpret and to bind to the original protocol.

During the last year, I worked on applying formal methods to security protocols with good results on a complex case study, e.g. Z-Wave protocol. The obtained results [4] show that ASMs (Abstract State Machines) can be a suitable way to describe complex protocols even for designers and engineers because the notion of ASM resembles that of the Finite State Machine (FSM), which they commonly know. Moreover, the ASMETA framework [1] (a tool-set for modelling, validating and verifying ASMs), which is at the basis of my experimentation, is easily expandable with new external tools or software components.

The development of the desired modelling and verification framework for security protocols can be reached by addressing several research questions stated below and regarding three main precise research directions:

© Springer Nature Switzerland AG 2021
A. Raschke and D. Méry (Eds.): ABZ 2021, LNCS 12709, pp. 158–161, 2021.
https://doi.org/10.1007/978-3-030-77543-8_16

- *RQ1: "How is it possible to take advantage of all the pros of a formal method without the need to know its mathematical formalism?"*
  In particular, I am interested in evaluating a front-end interface that hides the mathematical formalism and using natural language processing (NLP) techniques that can automatize the extraction of requirements from the documentation [8].
- *RQ2: "Are classical model checkers suitable enough to guarantee properties of secure systems?"*
  Classical verification properties are qualitative and are related to functional behaviour. Nevertheless, security protocols operate in a hostile and unpredictable environment that requires the checking of non-functional properties. Therefore stochastic model checker should be tested to verify probabilistic and real-time behaviour.
- *RQ3: "How is it possible to guarantee that code operates as required by the specification?"*
  This is a fundamental question since the availability of these techniques may influence also the protocol certification process. To answer this question three possible directions can be investigated: code generation from models, conformance checking and runtime verification.

2 Related Work

In the last few years, many techniques and tools have been proposed for verifying protocols. The main points of difference between tools are the modelling language and the verification technique. In the rest of the section, I summarize the most relevant and mature verification tools.

TAMARIN [6] is a tool used to verify various security protocols, among which it is possible to find complex protocols like TLS, 5G and RFID. It has a modelling language that allows specifying protocols, adversary models and security properties. However, the verification phase is based on deduction and equational reasoning. The principal drawback of this tool is the need for human intervention during the verification phase.

ProVerif [3] uses pi-calculus as a modelling language, and for protocols verification, it checks the Horn clause. ProVerif may need human intervention when it may find a false attack or fail to prove some protocol properties.

Verifpal [9] is a relatively new framework inspired by ProVerif. It employs a language similar to the one used in protocol specification but with a fixed number of primitives. The verification phase relies on symbolic verification.

AVISPA [2] is a platform that group several tools, which use a single modelling language called HLPSL (High-Level Protocol Specification Language). The verification phase is handled using three different tools: SATMC (SAT-based Model-Checker) for abounded state space, CL-AtSe and OFMC for a bounded number of sessions, TA4SP for an unbounded number of sessions.

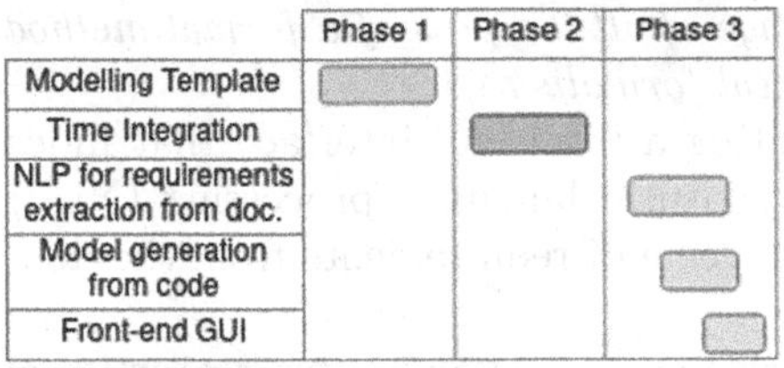

	Phase 1	Phase 2	Phase 3
Modelling Template	▭		
Time Integration		▭	
NLP for requirements extraction from doc.			▭
Model generation from code			▭
Front-end GUI			▭

	Phase 1	Phase 2	Phase 3
Verification Template	▭		
Flattener		▭	
Bounded Model Checker			▭
Stochastic Model Checker			▭
Trace Visualization			▭

Fig. 1. Roadmap of steps needed for the modelling phase

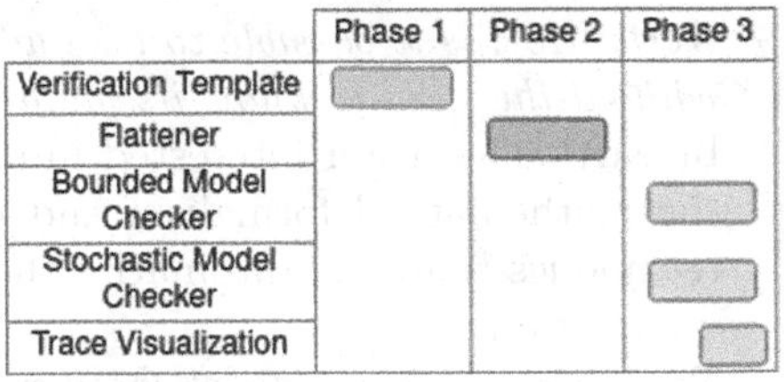

Fig. 2. Roadmap of steps needed for the verification phase

3   Description of the Approach

My PhD project aims to investigate new techniques to automatize both the procedure of modelling and verification, creating a tool that is user-friendly to spread the adoption of model checking in the industry sector.

Figure 1 shows the roadmap of steps needed to build a user-friendly and automated modelling tool for protocol security. The first step (the creation of a reusable template) [4] is the basis for all the future phases. It consists of three main sub-steps: a formalization of cryptographic primitives common among protocols, a template that formalizes agents' internal structure and the capability of interaction between agents using a message structure resembling the real-life equivalent. The second step in the roadmap plans to integrate time in the ASM formalism. The extension is needed because many security protocols are temporized. When the first two steps are completed can be used in the next steps. The template element will be mapped with a drag and drop graphical interface. More precisely, each construct available in the template is mapped with a UML-like element. The last two steps simplify the construction of the formal model, reducing designers intervention. The first one using NLP techniques can build a model from scratch extracting requirements from documentations. The second one is useful when implementation is provided by third-parties and it is necessary to guarantee that code operates in conformance with the specification. Therefore, I aim to automatically deriving the model from the code of the security protocol.

Figure 2 points out the steps needed for fully automated and exhaustive verification of security protocols. The first step that I have identified requires constructing a template of CTL properties reusable across protocols. A set of reusable security properties are presented in [3]. They cover the basic CIA (confidentiality, integrity and authenticity) properties.

The ASMETA framework already supports the flattening of the model in NuSMV [5], but it would be interesting to test other verification techniques (e.g. SAT solver) to compare performance and results. However, security protocols require the verification of qualitative properties, which means to verify properties that are, in many cases, probabilistic and real-time. The classical model checker can not deal with these properties, so stochastic model checkers will help execute an accurate verification.

The final goal is to map resulting traces with a graphical visualization similar to the one used during the model construction. In this way, designers or any user can easily spot the flaw.

4   Conclusion

Wider adoption of formal methods for security protocols design is crucial for obtaining more secure and robust protocols. In this paper, I outline my PhD research project, which aims to hide the mathematical complexities of formal methods by developing two main research areas, one centred on the modelling phase and the other on the verification. The main result that I plan to obtain is to spread the necessity of implementing security by design to reduce expensive flows and leaks during the protocol's life, preserving final users' privacy.

References

1. Arcaini, P., Gargantini, A., Riccobene, E., Scandurra, P.: A model-driven process for engineering a toolset for a formal method. Softw.: Pract. Experience **41**(2), 155–166 (2011)
2. Armando, A., et al.: The AVISPA tool for the automated validation of internet security protocols and applications. In: Etessami, K., Rajamani, S.K. (eds.) CAV 2005. LNCS, vol. 3576, pp. 281–285. Springer, Heidelberg (2005). https://doi.org/10.1007/11513988_27
3. Blanchet, B.: An efficient cryptographic protocol verifier based on prolog rules. In: Proceedings 14th IEEE Computer Security Foundations Workshop, 2001, pp. 82–96 (2001). https://doi.org/10.1109/CSFW.2001.930138
4. Braghin, C., Lilli, M., Riccobene, E.: Towards ASM-based automated formal verification of security protocols. In: Rigorous State-Based Methods. Springer (2021). (accepted)
5. Cimatti, A., et al.: NuSMV 2: an OpenSource tool for symbolic model checking. In: Brinksma, E., Larsen, K.G. (eds.) CAV 2002. LNCS, vol. 2404, pp. 359–364. Springer, Heidelberg (2002). https://doi.org/10.1007/3-540-45657-0_29
6. Cortier, V., Delaune, S., Dreier, J.: Automatic generation of sources lemmas in TAMARIN: towards automatic proofs of security protocols. In: Chen, L., Li, N., Liang, K., Schneider, S. (eds.) ESORICS 2020. LNCS, vol. 12309, pp. 3–22. Springer, Cham (2020). https://doi.org/10.1007/978-3-030-59013-0_1
7. Davis, J.A., et al.: Study on the barriers to the industrial adoption of formal methods. In: Pecheur, C., Dierkes, M. (eds.) FMICS 2013. LNCS, vol. 8187, pp. 63–77. Springer, Heidelberg (2013). https://doi.org/10.1007/978-3-642-41010-9_5
8. Ghosh, S., Elenius, D., Li, W., Lincoln, P., Shankar, N., Steiner, W.: Automatically extracting requirements specifications from natural language, March 2014
9. Kobeissi, N., Nicolas, G., Tiwari, M.: Verifpal: cryptographic protocol analysis for the real world. In: Bhargavan, K., Oswald, E., Prabhakaran, M. (eds.) INDOCRYPT 2020. LNCS, vol. 12578, pp. 151–202. Springer, Cham (2020). https://doi.org/10.1007/978-3-030-65277-7_8

Formalizing the Institution for Event-B in the Coq Proof Assistant

Conor Reynolds[(✉)] [iD]

Maynooth University, Maynooth, Kildare, Ireland
conor.reynolds@mu.ie

Abstract. We formalize a fragment of the theory of institutions sufficient to establish basic facts about the institution EVT for Event-B, and
its relationship with the institution $FOPEQ$ for first-order predicate logic.
We prove the satisfaction condition for EVT and encode the institution
comorphism $FOPEQ \to EVT$ embedding $FOPEQ$ in EVT.

Keywords: Coq · Event-B · Institution theory

1 Introduction

The theory of institutions [4] was introduced by Joseph Goguen and Rod Burstall
to give concrete form to the informal notion of a "logical system", identifying a
common structure among the many logics in regular use in computer science. A
2017 paper by Marie Farrell, Rosemary Monahan, and James Power [3] uses the
theory of institutions to provide a sound mathematical semantics and modular-
ization constructs for the industrial-strength state-based formal modelling language Event-B [1], providing interoperability with other formalisms. In related
work, the Heterogeneous Tool Set (Hets) [7] makes use of institutions to provide
heterogeneous specifications.

Event-B has an associated development process for system-level modelling
and analysis. Key features include the use of set theory as a modelling notation,
the use of refinement to represent systems at different abstraction levels and the
use of mathematical proof to verify consistency between refinement levels. The
primary purpose of this research is to formalize the work in [3] within the Coq
proof assistant, and more generally to provide the rudiments of a Coq library
for the theory of institutions.

We build on earlier work formalizing universal algebra in Agda by Emmanuel
Gunther, Alejandro Gadea, and Miguel Pagano [5]. However, the purpose of this
work is not to provide a comprehensive development of universal algebra; we only
develop as much as we need in order to define the institutions for first-order logic
and Event-B. We also depend on the development of category theory by John
Wiegley at jwiegley/category-theory.

Supported by the Irish Research Council (GOIPG/2019/4529).

A. Raschke and D. Méry (Eds.): ABZ 2021, LNCS 12709, pp. 162–166, 2021.
https://doi.org/10.1007/978-3-030-77543-8_17

While some obligations remain to be formally discharged for the institution *FOPEQ* for first-order predicate logic with equality, our developments for the institution *EVT* for Event-B are complete. We have also encoded the institution comorphism *FOPEQ* → *EVT*, which embeds the simpler *FOPEQ* institution into *EVT*, providing the underlying mathematical language for *EVT*. It remains, however, to prove the naturality condition in our encoding. The formalization is not axiom-free, assuming dependent function extensionality and proof irrelevance. A more careful development might use setoids (as in [2,5]), and in the future we may experiment with grounding these efforts in homotopy type theory.

Throughout this paper, we will assume some familiarity with basic category theory, as well as the first two chapters of [8].

2   The Institution for Event-B

An *institution* [4] consists of

- a category Sig of signatures (non-logical syntax);
- a sentence functor Sen: Sig → Set (logical syntax);
- a model functor Mod: Sig^{op} → Cat (semantics for non-logical syntax); and
- a semantic entailment relation $\vDash_\Sigma \subseteq |\mathsf{Mod}(\Sigma)| \times \mathsf{Sen}(\Sigma)$ for each $\Sigma \in$ Sig,

such that for any signature translation $\sigma\colon \Sigma \to \Sigma'$, any sentence $\phi \in \mathsf{Sen}(\Sigma')$, and any model $M' \in \mathsf{Mod}(\Sigma')$, the satisfaction condition holds:

$$M' \vDash_{\Sigma'} \mathsf{Sen}(\sigma)(\phi) \quad \text{iff} \quad \mathsf{Mod}(\sigma)(M') \vDash_\Sigma \phi \tag{1}$$

This kind of institution is sometimes referred to as a set/cat institution, since the target of Sen is Set and the target of Mod is Cat. To avoid encoding a "category of categories" in Coq, we implement set/set institutions [6].

We will now provide a precise but brief definition for the institution for Event-B, alongside its definition in Coq. For details, we refer the reader to [3]. Throughout, let Status = {ordinary ≤ anticipated ≤ convergent}.

The category of *EVT*-signatures has as objects $\hat{\Sigma} = \langle \Sigma, E, X, X' \rangle$, where Σ is a first-order signature, $E :$ Status → Type is a status-indexed set of events, and $X, X' :$ sorts Σ → Type are sorts-indexed sets of pre- and post-variables, respectively. In Coq, this becomes:

```
Record EvtSignature :=
  { base_sig :> FOSig ;
    events : Status → Type ;
    Vars   : sorts base_sig → Type ;
    Vars'  : sorts base_sig → Type }.
```

An *EVT*-signature morphism $\hat{\Sigma}_1 \to \hat{\Sigma}_2$ consists of a first-order signature morphism $\sigma\colon \Sigma_1 \to \Sigma_2$ translating the base signature, along with a function $E_1 \to E_2$ mapping events in such a way as to preserve the ordering on statuses, and functions $X_1 \to X_2 \circ \sigma$, $X'_1 \to X'_2 \circ \sigma$ mapping variables, regarded as morphisms in their respective indexed categories. It is convenient to assume that the initialization event is not in E, so there is no need for the assumption that the initial event is preserved by signature morphisms. If the initialization/event distinction is made at the level of sentences, then we can enforce preservation of the initialization event definitionally.

```
Record EvtSigMorphism Σ Σ' : Type :=
  { on_base_sig :> SignatureMorphism Σ Σ' ;
    on_events : EventMorphism Σ Σ' ;
    on_vars   : Vars Σ → Vars Σ' ∘ on_base_sig ;
    on_vars'  : Vars' Σ → Vars' Σ' ∘ on_base_sig }.
```

EVT-sentences are either *initialization* sentences, Init ψ where ψ : FOSen($\Sigma + X'$), or *event* sentences, Event e ψ where ψ : FOSen($\Sigma + X + X'$). Note that the base signature is expanded to include the EVT-variables as constant operation names. Initialization sentences describe how variables are initially set. Event sentences describe how events change the variables. As a very simple example, given an event inc which increments a variable n, inc $:\equiv$ **begin** $n := n + 1$ **end**, we write the EVT-sentence Event(inc, $n' = n + 1$), where $n \in X$ and $n' \in X'$ are respectively pre- and post-variables from the ambient Event-B signature. Given an initialization event which starts n at 0, init $:\equiv$ **begin** $n := 0$ **end**, we write the EVT-sentence Init($n' = 0$). For details on this correspondence, see again [3].

Event-B sentences rely on the ability to construct the expansion of first-order signatures by adjoining a sorts-indexed set of constant operation names, which in Coq we denote by SigExpand Σ X. EVT-sentences can be defined as follows.

```
Inductive EVT Σ : Type :=
  | Init  : FOSen (SigExpand Σ (Vars' Σ)) → EVT Σ
  | Event : ∀ status, events Σ status
          → FOSen (SigExpand Σ (Vars Σ + Vars' Σ))
          → EVT Σ.
```

An EVT-model consists of a first-order model M and a pair of environments $L : \text{List}(X' \to M)$ and $R : E \to \text{List}(X + X' \to M)$, which are lists of valuations of variables in M. We enforce that L and R_e, for each event e, are nonempty.

```
Record EvtModel Σ :=
  { base_alg :> Algebra Σ ;
    envL : NEList (Vars' Σ → base_alg) ;
    envR : ∀ status,
           events Σ status → NEList (Vars Σ + Vars' Σ → base_alg) }.
```

Let M^θ denote the expansion of a model M by a valuation $\theta : X \to M$. We say that $\langle M, L, R \rangle \models \text{Init } \psi$ if for all valuations $\theta \in L$, we have $M^\theta \models \psi$, and we say that $\langle M, L, R \rangle \models \text{Event } e \ \psi$ if for all valuations $\theta \in R_e$ we have $M^\theta \models \psi$. This can be written down directly in Coq.

```
Definition interp_evt Σ M φ : Prop :=
  match φ with
  | Init ψ    => List.Forall (λ θ, AlgExpansion M θ ⊨ ψ) (envL M)
  | Event e ψ => List.Forall (λ θ, AlgExpansion M θ ⊨ ψ) (envR M e)
  end.
```

Now, taking a top-down perspective, we can define institutions in Coq as follows:

```
Class Institution :=
  { Sig : Category ;
    Sen : Sig → SetCat ;
    Mod : Sig^op → SetCat ;
    interp : ∀ Σ : Sig, Mod Σ → Sen Σ → Prop ;

    satisfaction : ∀ (Σ Σ' : Sig) (σ : Σ → Σ')
                     (φ : Sen Σ) (M' : Mod Σ'),
      interp M' (fmap[Sen] σ φ) ↔ interp (fmap[Mod] σ M') φ }.
```

Proving that EVT is an institution amounts to instantiating this class to the above definitions and discharging the generated obligations. The proofs rely on custom induction principles for the dependent records we introduce above, since the induction principles generated by Coq are too strong. For example, if one wishes to prove that two Event-B signature morphisms $\hat{\sigma}$ and $\hat{\sigma}'$ are equal, of course it suffices to prove that they are equal componentwise. Consider equality on the on_vars component. The statement of this equality will depend on a proof $p : \sigma = \sigma'$ that the underlying first-order signature morphisms are equal, which we write $p_*(\text{on\_vars } \hat{\sigma}) = \text{on\_vars } \hat{\sigma}'$. Notice that this requirement is substantially stronger than necessary; it suffices in this case to know that σ and σ' agree on sorts. Hence, given $p' : \text{on\_sorts } \sigma = \text{on\_sorts } \sigma'$, we only need to prove $p'_*(\text{on\_vars } \hat{\sigma}) = \text{on\_vars } \hat{\sigma}'$. This dramatically simplifies the proofs.

3   Future Work

In the future, it will be interesting to investigate Coq's code extraction facilities to generate provably correct code derived from, for example, the institution comorphism $FOPEQ \to EVT$. We also wish to prove the amalgamation property for EVT, and more generally to build institution-independent constructions and proofs, which we have already explored to some extent for modal logics and linear-time temporal logics. The proofs involved in the definition for first-order predicate logic were rather complicated, but the proofs for EVT often reduced to properties of first-order logic. This suggests that quick progress could be made defining further institutions, verifying their properties, and providing interoperability between represented formalisms represented in our framework.

References

1. Abrial, J.R.: Modeling in Event-B: System and Software Engineering. Cambridge University Press, Cambridge (2010)
2. Capretta, V.: Universal algebra in type theory. In: Bertot, Y., Dowek, G., Théry, L., Hirschowitz, A., Paulin, C. (eds.) TPHOLs 1999. LNCS, vol. 1690, pp. 131–148. Springer, Heidelberg (1999). https://doi.org/10.1007/3-540-48256-3_10
3. Farrell, M., Monahan, R., Power, J.F.: An institution for Event-B. In: James, P., Roggenbach, M. (eds.) WADT 2016. LNCS, vol. 10644, pp. 104–119. Springer, Cham (2017). https://doi.org/10.1007/978-3-319-72044-9_8

4. Goguen, J.A., Burstall, R.M.: Institutions: abstract model theory for specification and programming. J. ACM **39**(1), 95–146 (1992)
5. Gunther, E., Gadea, A., Pagano, M.: Formalization of universal algebra in Agda. Electron. Notes Theor. Comput. Sci. **338**, 147–166 (2018)
6. Mossakowski, T., Goguen, J., Diaconescu, R., Tarlecki, A.: What is a logic? In: Beziau, J.Y. (ed.) Logica Universalis, pp. 111–133. Birkhäuser Basel (2007)
7. Mossakowski, T., Maeder, C., Lüttich, K.: The heterogeneous tool set, HETS. In: Grumberg, O., Huth, M. (eds.) TACAS 2007. LNCS, vol. 4424, pp. 519–522. Springer, Heidelberg (2007). https://doi.org/10.1007/978-3-540-71209-1_40
8. Sannella, D., Tarlecki, A.: Foundations of Algebraic Specification and Formal Software Development. Springer-Verlag (2012)

Author Index

Aniello, Leonardo   34

Baugh, John   99
Berglehner, Randolf   143
Bodeveix, Jean-Paul   66
Bombarda, Andrea   105
Bonfanti, Silvia   105
Braghin, Chiara   17
Butler, Michael   34, 124, 130

Coudert, Sophie   50

Dghaym, Dana   34, 124, 130
Du, Yiqing   3
Dyer, Tristan   99

Eschbach, Robert   137

Filali, Mamoun   66

Gargantini, Angelo   105
González, Senén   3, 118

He, Shilan   3
Hoang, Thai Son   34, 124, 130
Hu, Runshan   34

Jiang, Fengqing   118

Leuschel, Michael   81
Li, Zilinghan   3
Lian, Xinyu   118
Lilli, Mario   17, 158

Mashkoor, Atif   81
Moosbrugger, Jakob   112

Paulweber, Philipp   112

Rasheeq, Abdul   143
Reynolds, Conor   162
Riccobene, Elvinia   17, 105
Riviere, Peter   153

Salehi Fathabadi, Asieh   124, 130
Salunkhe, Shubhangi   143
Sassone, Vladimiro   34
Schewe, Klaus-Dieter   3, 118
Snook, Colin   124, 130

Vu, Fabian   81

Xiong, Neng   118

Zdun, Uwe   112

Printed in the United States
by Baker & Taylor Publisher Services